Oxford Chemistry Series

Oxford Chemistry Series

1972

1. K. A. McLauchlan: *Magnetic resonance*
2. J. Robbins: *Ions in solution* (2): *an introduction to electrochemistry*
3. R. J. Puddephatt: *The periodic table of the elements*
4. R. A. Jackson: *Mechanism: an introduction to the study of organic reactions*

1973

5. D. Whittaker: *Stereochemistry and mechanism*
6. G. Hughes: *Radiation chemistry*
7. G. Pass: *Ions in solution* (3): *inorganic properties*
8. E. B. Smith: *Basic chemical thermodynamics*
9. C. A. Coulson: *The shape and structure of molecules*
10. J. Wormald: *Diffraction methods*
11. J. Shorter: *Correlation analysis in organic chemistry: an introduction to linear free-energy relationships*
12. E. S. Stern (ed): *The chemist in industry* (1): *fine chemicals for polymers*
13. A. Earnshaw and T. J. Harrington: *The chemistry of the transition elements*

1974

14. W. J. Albery: *Electrode kinetics*
16. W. S. Fyfe: *Geochemistry*
17. E. S. Stern (ed): *The chemist in industry* (2): *human health and plant protection*
18. G. C. Bond: *Heterogeneous catalysis: principles and applications*
19. R. P. H. Gasser and W. G. Richards: *Entropy and energy levels*
20. D. J. Spedding: *Air pollution*
21. P. W. Atkins: *Quanta: a handbook of concepts*

W. S. FYFE
PROFESSOR OF GEOLOGY, UNIVERSITY OF WESTERN ONTARIO

Geochemistry

Clarendon Press · Oxford · 1974

Oxford University Press, Ely House, London W.1

GLASGOW NEW YORK TORONTO MELBOURNE WELLINGTON
CAPE TOWN IBADAN NAIROBI DAR ES SALAAM LUSAKA ADDIS ABABA
DELHI BOMBAY CALCUTTA MADRAS KARACHI LAHORE DACCA
KUALA LUMPUR SINGAPORE HONG KONG TOKYO

PAPERBACK ISBN 0 19 855429X
CASEBOUND ISBN 0 19 8554672

PRINTED IN GREAT BRITAIN BY
J. W. ARROWSMITH LTD., BRISTOL, ENGLAND

Editor's Foreword

FOR past students of chemistry, the word 'geochemistry' has been associated with mineralogy (thought to be dull) and silicate chemistry (known to involve odd names and unintelligible formulae). This book should certainly dispel these ideas. First of all, it demonstrates the vast scope of the subject, embracing all the cosmos, dealing with all the elements, exploiting all branches of chemistry, and extending chemistry into unfamiliar regions of high pressure and temperature. Secondly, it shows that geochemistry is not concerned with static systems; though rates of change may be slow, the changes themselves, in the earth's crust for example, are far-reaching in interest and importance. Thirdly, it shows that geochemistry is an infant science—so much is unknown and so much has to be inferred from so little evidence; the prospects for further research are immense. Not only are these prospects immense, but they now have urgency: man's ability to investigate (and therefore interfere with) what lies below his feet or stretches above his head is now so advanced that a real study of geochemical phenomena is both feasible and indeed essential for his long-term survival.

A.K.H.

Preface

EARTH chemistry, or geochemistry, is a subject with a history deeply rooted in the history of all physical and observational science. Observations on minerals and rocks and on the nature of objects in space intrigued the ancients just as they continue to fascinate us today. It has been said (*pace* my chemical friends) that geology and astronomy are the only two fundamental sciences. The history of mineralogy is part of the story of the discovery of the elements. The science of crystallography has provided the foundation to the physics of the solid state. Steno's discovery (1671) of the constancy of the angles between similar faces on crystals of the same substance was a vital contribution to atomic theory. Leonardo da Vinci and Robert Boyle were among the earliest oceanographers. Perhaps one of the great discoveries of this century has been that of finding that the earth today is in a state of vigorous convection (associated with so-called continental drift or ocean-floor spreading or plate tectonics).

We still know little about the way our earth functions. But today we live in a period of renaissance of the observational sciences. The population explosion and the side effects of pollution, resource development and conservation, and the like, are real. At the time of writing this book there is an energy crisis. But the crisis is not really one of energy, for such resources (e.g. solar energy) are enormous. The crisis lies only in our intelligence and wisdom. The conservation of our earth for man, the strange animal who demands both television sets and empty lands with forests and animals, presents a challenge on a scale we have never faced before. The only hope that we find solutions to such problems, that the renaissance is fruitful, must lie in the ability of all types of scientists to communicate. It is to such communication that this little book is dedicated.

Chairman Geology
University of Western Ontario
London, Canada

W. S. FYFE, F.R.S.

Acknowledgements

PERMISSION to reproduce the material listed below is gratefully acknowledged.

Figs. 2.2a, 5.1, 5.2: after Bullen, K. E., *An introduction to the theory of seismology*, 1963, Cambridge University Press.
Fig. 2.2b, after Verhoogen, *et al.*, *The Earth*, Holt, Rinehart and Winston Inc.
Fig. 3.6: after Fyfe, Geochemistry of Solids, © 1964, McGraw-Hill Inc.
Fig. 6.8: from Abelson, *Researches in geochemistry*, © 1959, John Wiley & Sons Inc.
Fig. 7.1: after Helegson, *Complexing and hydrothermal ore deposits*, 1964, Pergamon Press.
Fig. 7.3: after Barnes, *Geochemistry of hydrothermal ore deposits*; 1967, Holt, Rinehart and Winston Inc.
Fig. 8.1: from Craig, *The upper atmosphere and meteorology*, Academic Press.
Figs. 8.2, 8.3: after Valley, *Handbook of geophysics and space environments*, 1965, McGraw-Hill Book Co. Inc.
Fig. 9.1: from Gass, Smith, Wilson, *Understanding the Earth*, 1971, Artemis Press (for the Open University).

My thanks must also go to many colleagues and students who have done their best to keep me slightly mentally alive. And final thanks to my office staff for all their work and to all those in the Clarendon Press for their tolerance and assistance.

Contents

1. Introduction: what is geochemistry?

TRADITIONAL geochemistry involves the description of the chemistry of the earth. It attempts to describe the distribution of the elements and their isotopes in the various more-or-less well-defined parts of the earth; atmosphere, hydrosphere,*† crust,* mantle,* core.* Traditional geochemistry has also been concerned with obtaining similar data for extraterrestrial matter; meteorites, sun, stars. Geochemistry has evolved rather like inorganic chemistry, from a descriptive stage to one increasingly concerned with the mechanism behind the observed facts.

Modern geochemistry is concerned with the integration of chemical and geological approaches to the very general problem of how the earth and solar system have evolved over the period of $5{\cdot}0 \times 10^9$ years or so since their birth. Today, modern geological and geophysical research has demonstrated that the earth is in a convecting state. New materials arrive at the surface from the interior and at the same time parts of the surface return to the interior. Broadly, equilibrium is only local and the system is in an approach to a steady state. For example, most of the materials making the crust of the present ocean floors has been formed in the last 10^8 years or so. But it is also certain that the pattern of convection has changed through time and the present motions have changed in pattern and intensity.

There is probably no part of modern chemical science that does not have significant applications in earth science. The earth is a large object and in general most processes are slow. The time scale of many events is far removed from a laboratory or human time-scale. Most materials of the earth are subjected to, and change in response to pressures greater than, 100 kilobars and temperatures greater than 1000°C. Again, the conditions, and the materials involved, may not be very familiar to us. Often experiment is difficult and expensive. It is thus essential to understand the 'rules of the game' and as we must extrapolate and interpolate, the general ideas from thermodynamics and reaction-rate theory are vital.

To man, the surface environment has dominant importance. The growth of human population has forced us to appreciate that the earth is not infinitely tolerant to change on a human time-scale. We are now beginning to understand the interactions of the system I term 'air–life–water–rock'. Environmental chemistry is one of the most intricate problems that man must attempt to understand during the next decades. It involves complex equilibrium and rate aspects of a heterogeneous gas–liquid–solid system strongly influenced by

† Terms marked thus are explained in the Glossary (p. 103).

biological activity. When we look at the earth's surface we tend to underestimate the latter, but some mountain ranges are formed by materials synthesized in living cells; some of the most important parts of ocean–atmosphere chemistry are regulated by biological activity and our coal–gas–petroleum resources depend on past biological activity. The problem of understanding and hence regulating or modifying the balance is tremendously difficult. There is no easy path to such an understanding but the geological and chemical sciences are at the core of any realistic study.

Man makes a steadily increasing demand on raw materials, and the problem of maintaining the supply becomes steadily more difficult. In the final analysis, rocks provide these needs whether it be iron ore for steel or phosphate for fertilizers. The formation of an ore deposit normally involves some form of transport and concentration. Natural purification processes are often extremely efficient (again a function of time and scale) but in general the chemistry behind them is only guessed at. Improved understanding of the chemistry of transport processes in the earth is vital if we are to utilize and conserve raw materials. Essential to the understanding of such processes is a more exact knowledge of the molecular chemistry of inorganic fluids to 1000°C or so.

When geochemical processes are introduced to the student of chemistry it is often necessary to stress that geochemical scales are highly varied both in time and mass (as well as P and T). Scale often imposes constraints on what can and cannot happen. At the present time silicate liquids from depth arrive at the surface at rates of about 10^{14} kg per year. A single bubble of melt rising into the crust may be 500 km^3 in volume. It may heat several thousand km^3 of sea water to elevated temperatures and the fluids may extract and precipitate metal-rich phases over a path-length of kilometers. The melt bubble may take more than a million years to cool. Geochemical 'beakers' are often large and natural phenomena have large time constants. At the same time, the micro-mechanism involves the time-constants of electron–atomic motion.

In summary, there are few aspects of earth science that do not require the knowledge and skills of the chemist in their study and there are few parts of earth chemistry that, when adequately studied, will not yield new concepts for chemistry and often useful ideas for man. In a book of this length one can only hope to provide a very slight introduction to the types of problem that concern the geochemist.

2. Some basic data about the earth

AT the outset, we may consider some basic data about this planet which eventually must fit into the geochemical history.

Mass

From the laws of gravitation, the determination of the gravitational constant, and the acceleration due to gravity, and from the motion of the moon and satellites, the mass of the earth is determined as $5{\cdot}976 \times 10^{27}$ g.

Radius

The mean radius of the earth is 6371 km but the earth is not quite a perfect sphere (equatorial radius 6378 km, polar radius 6356 km). The shape is a result of the manner of rotation and non-uniform mass distribution. From the mean radius and mass, the mean density of the earth is $5{\cdot}517$ g cm^{-3}. This is an important result because it is much larger than the observed densities of surface rock materials ($2{\cdot}5$–3 g cm^{-3}) and hence the interior must be denser due to either compression, phase changes, or changes in chemical composition.

Surface

The surface area of the earth is $5{\cdot}1 \times 10^8$ km^2. Of this $1{\cdot}49 \times 10^8$ km^2 is land surface and $3{\cdot}61 \times 10^8$ km^3 is ocean. As we shall see later, the chemical composition of the ocean floor is in general quite different from that of the continents. The mean elevation of the land surface is $0{\cdot}875$ km and the mean depth of the oceans $3{\cdot}8$ km. Neither the maximum elevation of mountains above sea level nor the maximum depth to the sea-floor below sea level exceeds 10 km. These numbers tell us something about the ability of the surface to carry loads or the strength of the surface layers.

Mass distribution

Many types of observation on the earth's gravitational field indicate that the earth is not quite in gravitational equilibrium and that mass is not uniformly distributed. In recent years an increasingly precise picture of mass distribution has come from data on satellite motions. It is obvious that if the earth was in a state of perfect gravitational equilibrium, mass would be distributed out from the centre in a radially symmetrical pattern. There might be small perturbations resulting from the pattern of earth rotation but a satellite would follow some perfectly elliptical trajectory. This is found not to be the case and there are departures from the equilibrium figure. Similar departures have been found on the moon (mascons*). Again a departure from the equilibrium state means either that the earth has considerable strength or that convective motions are active. As we proceed we shall see convincing evidence for the latter.

Significant data also come from the study of the moment of inertia. The moment of inertia of a body about any axis is related to the distribution of mass about that axis. The earth's moments of inertia do not correspond to a uniform mass distribution through the earth but rather a large concentration of mass towards the centre. This is one of the important facts that must be considered in producing a density model for the earth.

Magnetic field

We know the earth has a magnetic field. We know also that this field is subject to reversals of polarity. Over the past three million years the polarity has reversed about 20 times. This does not seem typical of all geologic time for there can be long periods of stability. But there must be some process, a rather dynamic one, that produces the magnetic field. Theories of the origin and mobile properties of the field are imperfect, but in our present state of knowledge it is considered to be a magnetohydrodynamic result of motions in a metallic (conducting) liquid which, considering the mean density and moment of inertia, could be a liquid metal like iron.

Age

We do not know the age of the earth. From applications of the decay laws and decay constants of radioactive isotopes we can find the oldest rocks. Naturally as with time more rocks are sampled, the maximum age of the surface increases. At the time of writing this book the oldest rocks found on earth come from Greenland and are about $3{\cdot}9 \times 10^9$ years old. The oldest rocks tend to be concentrated in the interiors of the major continents.

Studies of lunar materials and meteorites, indicate that solid objects existed in the solar system about $4{\cdot}5 \times 10^9$ years ago. Estimates of age (or intelligent guesses) based on present lead isotope ratios and knowledge of thorium–uranium decay would place an age limit of about 5·5 billion* years on terrestrial ages. It has also been shown that certain anomalous isotopes in meteorites (e.g. ^{129}Xe derived from ^{129}I, half-life 17 My) indicate a very short period between some nucleosynthesis and accumulation. All things considered, the solar system seems to be about 5 billion years old.

The oldest rocks on earth come from continental masses. A fascinating discovery of the past decades is that the rocks of the ocean floors are all quite young. Most rocks on the ocean floors are younger than 200 million years. These age observations must be fitted to our earth model but they clearly suggest that recycling is occurring; or perhaps we could call it convection.

Composition of the observable region

The only parts of the earth whose chemistry can be adequately sampled involve: atmosphere, hydrosphere, and crust. The deepest mines and bore-holes probe the top 8 km or so. As we go beyond, speculation increases. We shall return to this subject in Chapter 5 but suffice to say here that 99 per cent of the solid crust is made of eight elements (O, Si, Al, Fe, Ca, Na, K, Mg in order of abundance). These elements form a rather simple array of low-density minerals. Crustal element abundances do not reflect solar or universal abundances and do not easily fit the major geophysical parameters of the earth, and hence we can be certain that the crust does not represent the composition of the entire earth. Thus we are faced with one of the greatest problems in geochemistry. What is the composition of the entire earth?

Heat flow

If we drill a hole into the crust we find that the temperature rises by about 20–30°C for every kilometer increase in depth. Silicate melts appear at the surface of the earth during volcanic eruptions at temperatures up to 1200°C or so. We can be certain that the interior is hot. This implies that heat flows continuously to the surface. We can measure the thermal gradient in the crust and the thermal conductivity of rocks, and hence this heat flow. Heat flow is variable, there are hot spots and cold spots, but in general values around 40–80 $mJ\ m^{-2}\ s^{-1}$ are found. Solar energy controls the temperature in the upper few metres but most of this energy is returned to space.

There are a limited number of heat sources for the earth such as: heat from release of gravitational energy during accumulation and redistribution of mass, heat from tidal friction caused by the influence of sun and moon, heat from radioactivity. It is the latter term that involves geochemicsts for it requires a knowledge of the distribution of heat-producing elements (uranium, thorium, potassium-40) in the earth. Heat production sets definite limits on what compositions are possible at depth. The crust is much too radioactive to be representative of the whole earth.

If the thermal gradient of 20–30°C km^{-1} observed at the surface was maintained to great depths, the temperature at the centre would exceed 120 000°C. Experimental studies show that no known substances could be solid at such a temperature and the pressures at the core. Thus thermal gradients must become much less at depth within the earth. The surface variation in heat flow (and naturally the occurrences of things like volcanoes) shows that either heat sources are not uniformly distributed or that the earth is in some form of convective motion.

Seismic wave propagation

When a sudden stress (explosion, earthquake, etc.) is applied to a solid body, vibrational waves are propagated away from the sources. From the theory of elasticity these waves are of two types:

S (shear) waves are propagated with a velocity

$$V_s = \left(\frac{\mu}{\rho}\right)^{\frac{1}{2}},$$

where ρ is the density and μ is the rigidity or shear modulus. S waves are not transmitted by normal liquids.

P waves have a velocity

$$V_p = \left(\frac{K+\frac{4}{3}\mu}{\rho}\right)^{\frac{1}{2}}$$

where K is the bulk modulus. P waves are transmitted by liquids.

These waves obey the normal laws of reflection and refraction and provide perhaps the most important tool for probing beneath the surface of the earth.

Major shocks (earthquakes, atomic explosions) are frequent. Seismograph* arrays are located over much of the earth and the passage of seismic waves generated in many depth regions can be studied with immense sophistication. Certain extremely valuable results are now established.

(1) The velocity of waves depends on the chemical composition and phase-state of a material. It has been shown experimentally that the mean atomic number of the material is critical.

(2) The low-density crust is about 20–70 km thick under the continents and 6–8 km thick under the oceans. Below this crust–mantle junction (the Moho*), seismic velocities increase sharply.

(3) In general wave velocity increases with depth, but not at a steady rate. At the 100–200 km level there appears to be a zone of lower velocities, which may result from a small degree of melting.

(4) There appear to be zones at 400 and 600 km depth where velocities increase rapidly and phase changes are suggested.

(5) At 2900 km there is a drastic change. V_s drops to zero suggesting a liquid state and V_p drops by about 40 per cent.

(6) At even greater depth, V_p increases in a central solid core of about 1200 km radius which is itself probably layered, indicating either that a phase change has occurred in the core material or that the liquid–solid core represents crystal–liquid equilibrium in a polycomponent system.

These observations, now well established, must be seriously considered in any earth model. As we shall see later, seismic velocities also allow us to place limits on the type of materials which can exist at any depth in the earth. We can then assume an earth model as indicated in Fig. 2.1. The P-wave velocity profile is indicated in Fig. 2.2.

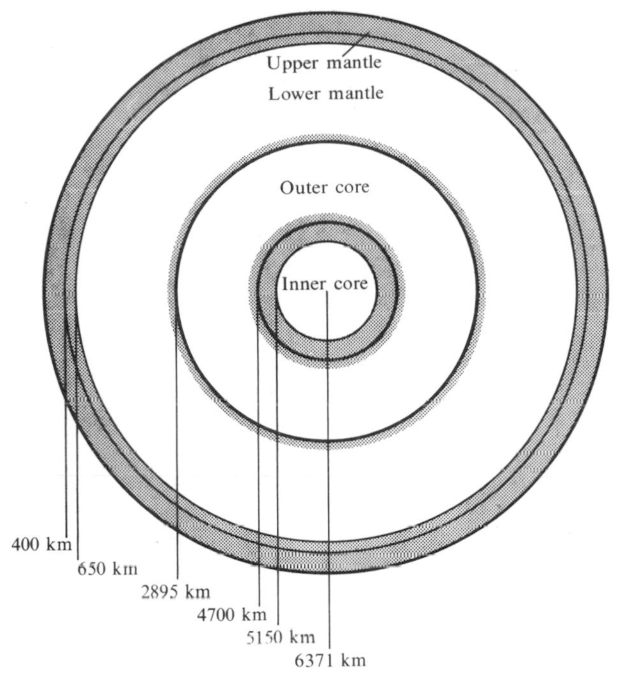

FIG. 2.1. General structure of the earth. The inner core is solid and layered, the outer core is liquid. There are discontinuities in the upper mantle (light lines) which probably represent regions of phase changes.

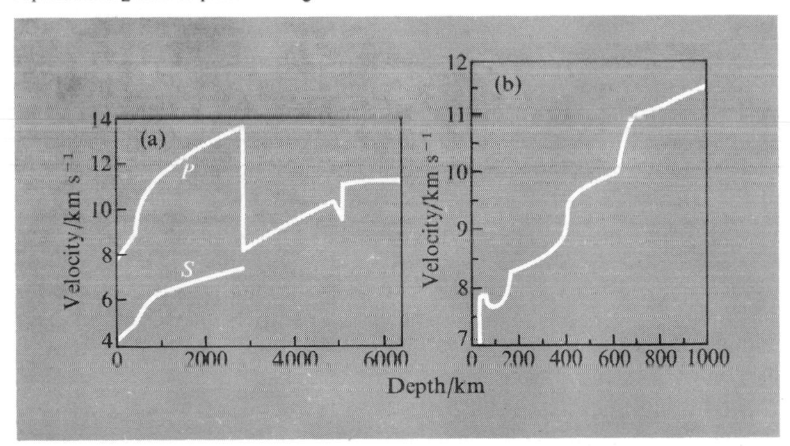

FIG. 2.2.(a) General pattern of P- and S-wave velocities in the earth. (b) Detailed P-wave profile for the upper 1000 km. Note the low-velocity region around 100 km which is normally attributed to partial melting. Other discontinuities are more likely to represent phase changes.

3. The composition of the crust: minerals

As mentioned on p. 6, evidence from seismology shows that the crust, with an average density of 2·8 g cm^{-3}, has a thickness of about 30 km in continental regions and 8 km under the oceans. Most of this crust can be sampled by bore holes or when, because of tectonic* processes (e.g. mountain building), deep material arrives in our field of observation. How do we obtain an average composition? One could imagine an experiment where holes are drilled on some statistical pattern and the materials chemically analyzed. But deep holes are expensive and difficult to drill, particularly in the oceans, and what spacing would be adequate? Realistically, it turns out that an intelligent averaging process must be based on geologic maps. If we can map the major rock-types of the crust (there are not more than a dozen) we can greatly reduce, and improve the sampling problem. By using such approaches the data indicated in Table 3.1 are produced. It should be noted that for many elements these figures represent only a good order-of-magnitude estimate.

Inspection of such data shows that every element (and isotope) that would be expected from our knowledge of nuclear stability and radioactive decay constants is present, even if in minor amounts. But it is also obvious that a few elements dominate the crust and it is no surprise that silicate minerals are the dominant type of compound. It is also obvious that some materials which we consider common in every day life are present in very small amounts (Cu, Pb, Sn). These elements become available in ores due to the quite remarkable efficiency of various natural concentration processes. If one compares data in Table 3.1 with cosmic abundances (p. 37) it is also clear that the crust is quite non-typical (e.g. compare K, U, Au). Our crust is obviously a highly-fractionated cosmic sample.

A word here is perhaps necessary on the analytical techniques used in determination of the chemistry of a rock or mineral. It is obviously a formidable task, particularly if data are required on a wide range of elements down to the parts-per-billion* level. Classical methods of analysis were based on essentially two techniques depending on the concentration level of the elements determined. For major constituents a system of gravimetric analysis was involved following solution of the material in either hydrofluoric acid or solution by fusion in a suitable flux. For trace elements, analysis of arc spectra has been the traditional method. Both techniques are still widely used, but during the past decades, many new and faster instrumental techniques have come into common use and have resulted in more voluminous and in some cases, more reliable data.

TABLE 3.1

Composition of the earth's crust expressed in parts per million

H	1520	Cu	68	Pr	9·1
Li	18	Zn	76	Nd	39·6
Be	2	Ga	19	Sm	7·0
B	9	Ge	1·5	Eu	2·1
C	180	As	1·8	Gd	6·1
N	19	Se	0·05	Tb	1·18
O	456 000	Br	2·5	Ho	1·26
F	544	Rb	78	Er	3·46
Na	22 700	Sr	384	Tm	0·5
Mg	27 640	Y	31	Yb	3·1
Al	83 600	Zr	162	Hf	2·8
Si	273 000	Nb	20	Ta	1·7
P	1120	Mo	1·2	W	1·2
S	340	Pd	0·015	Re	0·0007
Cl	126	Ag	0·08	Os	0·005
K	18 400	Cd	0·16	Ir	0·001
Ca	46 600	In	0·24	Pt	0·01
Sc	25	Sn	2·1	Au	0·004
Ti	6320	Sb	0·2	Hg	0·08
V	136	I	0·46	Tl	0·7
Cr	122	Cs	2·6	Pb	13·0
Mn	1060	Ba	390	Bi	0·008
Fe	62 200	La	34·6	Th	8·1
Co	29	Ce	66·4	U	2·3
Ni	99				

Data from Fairbridge (1972).

Some of the more important of such techniques include:

(1) X-ray fluorescence spectrography. In this technique characteristic X-ray spectra of the elements is produced by X-ray excitation. The X-ray spectrum so produced is analysed by using crystals of standard lattice spacing as a diffraction grating, essentially a result of Bragg's law for X-ray diffraction.

(2) Characteristic atomic X-ray spectra can also be generated by electron-bombardment. In the electron microprobe, a highly-focussed electron beam interacts with a small surface volume of a sample (10^{-18} m^3 or less) and again excites X-radiation which can be analysed as above. The method has wide sensitivity and is particularly useful for studying the chemistry of single crystals or small objects in general.

(3) Other analytical techniques depend on excitation or absorption of radiation of longer wavelengths. These include analysis of flame spectra in flame photometry or atomic absorption and various techniques for colorimetric analysis. Today, atomic absorption is proving to be one of the universal work-horses of geochemistry. Many elements can be determined, often at the p.p.m. or p.p.b. level.

(4) For the precise determination of elements at very low levels of concentration, activation analysis can be most valuable. In this technique, the sample is irradiated by a source of multienergy neutrons which form unstable nuclides of most elements. The decay pattern of each nuclide is different and by analysis of the spectrum of energetic decay species (electrons, γ-rays etc.) the presence and quantity of a given nuclide can be determined.

(5) Mass spectrographic determinations of isotope abundances have reached a remarkable degree of sensitivity. Such data are required in studies of stable isotopes and their fractionation and obviously are needed in work involving age determination. The modern mass spectrograph is also used for analytical purposes and on occasion can be highly versatile and superior to optical analysis of arc spectra. In such techniques, ions are separated on the basis of their charge–mass ratios, and modern electronic counting devices are of remarkable sensitivity.

In summary, a modern geochemical laboratory can determine most elements and isotopes at the levels they occur in most common earth materials. Frequently, the greatest practical problem is in the initial sampling of earth materials. It perhaps should be noted that the standard equipment for any such laboratory with wide ranging capabilities, will cost around a million dollars. Environmental chemistry is not expensive!

Minerals

Given a crust dominated by O, Si, Al, Fe, Ca, Na, K, Mg it is obvious that the most common minerals will be compounds of oxygen. In general, silicates and aluminosilicates make up the bulk of most rocks. From the above elements some of the very common rock forming minerals listed below are formed:

Oxides	SiO_2	quartz
	Fe_3O_4	magnetite
Feldspars	$NaAlSi_3O_8$	albite
	$KAlSi_3O_8$	orthoclase
	$CaAl_2Si_2O_8$	anorthite
Pyroxenes	$CaMgSi_2O_6$	diopside
	$MgSiO_3$	enstatite
Olivines	Mg_2SiO_4	forsterite
	Fe_2SiO_4	fayalite
Micas	$KAl_2AlSi_3O_{10}(OH)_2$	muscovite
	$KMg_3AlSi_3O_{10}(OH)_2$	phlogopite
Amphiboles	$Mg_7Si_8O_{22}(OH)_2$	anthophyllite
	$Ca_2Mg_5Si_8O_{22}(OH)_2$	actinolite

Between them such mineral types probably make up about 80 per cent or so of the crust. But the number of minerals is very large indeed. There are three major reasons for this. First, the range of conditions under which minerals are formed in terms of *P–T* variation is very large indeed. Thus phase changes abound. For example SiO_2 may exist in many forms: quartz, tridymite, cristobalite, coesite, stishovite; and the first three all have two crystallographic modifications (α and β). Thus in the system SiO_2 there are at least eight important modifications (see pp. 45–46).

Second, minerals crystallize in chemically-complex systems and solid solutions abound. Minerals tend to be impure. For example, Mg_2SiO_4 (forsterite) and Fe_2SiO_4 (fayalite) crystallize in identical structures. The Mg^{2+} and Fe^{2+} ions are similar in size, and normally when minerals of this type form, a solid solution $(Mg,Fe)_2SiO_4$ results which is the mineral olivine. In general, if end-member* minerals (pure compounds) have similar crystal structures and are comprised of atoms of similar size and if the binding forces are similar (i.e. covalent or ionic) they will tend to mix. This leads to immense chemical complexity in minerals. In a chemical sense, nearly every mineral grain is unique, and it was only when minerals were considered from a structural point of view that a rational and useful classification resulted.

Third, natural fractionation processes can be very efficient. Again this is a function of natural scales; a transport experiment may occur along a diffusion path miles long with a thermal gradient of only a few degrees. Thus while the bulk chemistry of a natural system may be simple, the actual chemistry at some micro-locality may be highly variable, and at such a site a mineral of some rare element may be formed.

General features of the atomic structure of minerals

Just as in pure chemistry, much of our geochemical thinking is based on models of chemical bonding but with particular emphasis on bonding in crystals. Crystal morphology or shape generally reflects internal atomic structure and is important in identifying minerals. The arrangement of atoms in a crystal can normally be correlated with models of predominant ionic or covalent bonding and which type of bonding is most likely is readily ascertained by consideration of the electronegativity differences between bonded atoms. At the outset one should emphasize however, that in crystals of multi-element compounds like most minerals, the 'prediction' of a crystal structure is about impossible and as in all structural chemistry, we would be in a sorry state without X-ray crystal-structure determination.

When bonded atoms are of similar or identical electronegativities, covalent bonds, with or without delocalized electrons (metals, graphite) will result and the crystal structure will be determined by the number of valence electrons and the types of orbitals (or hybrids) involved. The situation is simply illustrated by the modifications of carbon, graphite and diamond. In diamond, the $2s^22p^2$

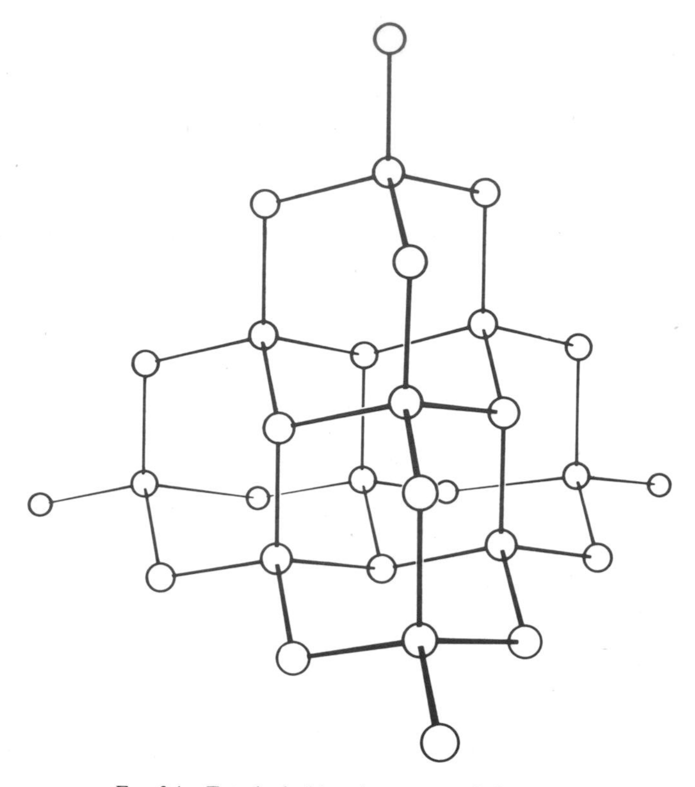

FIG. 3.1a. Tetrahedral bond structure of diamond.

electrons are involved in tetrahedral sp^3-hybrid orbitals to build the extremely strong three-dimensional structure of diamond (Fig. 3.1a). All electrons are used in localized bonds so that diamond is a perfect insulator. With graphite (Fig. 3.1b), the bonding involves an sp^2-trigonal planar carbon atom (as in ethylene) and leads to the formation of a two-dimensional polymer, the bonds being reinforced by a delocalized π-electron cloud formed by the remaining p-electrons. Graphite is an electrical conductor. Bonding between the carbon sheets in graphite involves only weak van der Waals forces. If an atom has few valence electrons, molecular compounds will result; the molecules may be linked by van der Waals forces or in appropriate cases by hydrogen bonds. Solids like sulphur, ice, solid helium, etc. exhibit these effects.

But most common minerals contain atoms like Mg or Na linked to oxygen and due to substantial electronegativity differences between bonded atoms one might expect that an ionic model could be more appropriate. The ionic

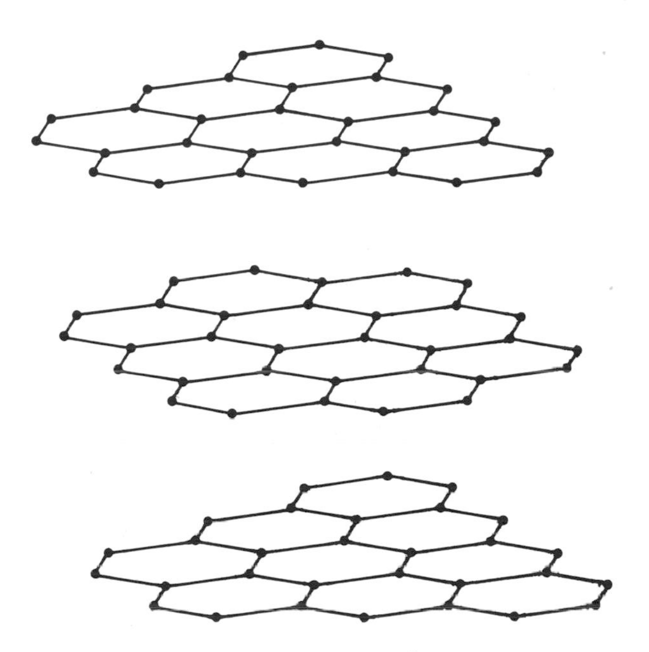

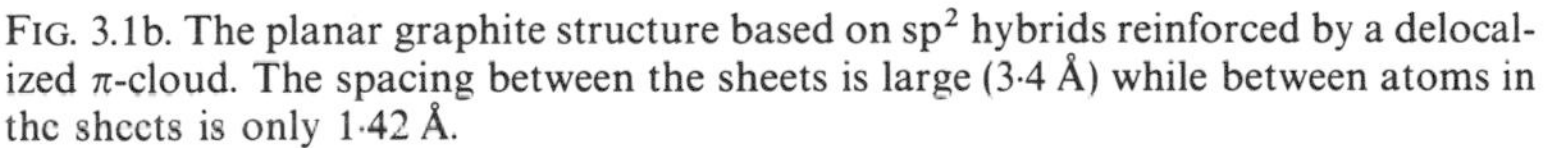

FIG. 3.1b. The planar graphite structure based on sp^2 hybrids reinforced by a delocalized π-cloud. The spacing between the sheets is large (3·4 Å) while between atoms in the sheets is only 1·42 Å.

model is quite different to a covalent model in its structural requirements. No geometric restrictions imposed by overlapping of atomic orbitals are involved; the ions are simply considered as structureless charged spheres. The behaviour of ionic crystals can be considered from the empirical equations for the energy of interaction of ions of charge z_1 and z_2 at a distance r. The equation for the potential energy V:

$$V = -\frac{z_1 z_2 e^2}{r} + \frac{be^2}{r^n}$$

indicates an energy of attraction obeying an inverse distance law, and a repulsion dependent on r^n; n is found to be of the order of 8–10 (b is a constant). This repulsion is 'hard' and has small values at large distances but becomes very large as r diminishes. It implies that ions have size and will not penetrate beyond a certain distance.

To obtain the energy of a crystal we must sum all such interactions:

$$V = \frac{-Az_1 z_2 e^2}{R} + \frac{be^2}{R^n}$$

where A is a summation constant (the Madelung constant) dependent on the crystal structure. At equilibrium:

$$\frac{dV}{dR} = 0$$

and the crystal energy is:

$$V = \frac{-Ae^2 z_1 z_2}{R}\left(1 - \frac{1}{n}\right).$$

The lattice energy U of a solid is usually defined as $-V$ and is the energy required to separate the crystal into ions at infinite separation.

These equations tell us many things about the ideal ionic crystal. First, ions have size. Second, stable structures will result from the maximum possible interaction of ions of opposite sign i.e. maximum coordination numbers possible within the restrictions of size and overall neutrality of the crystal.

A table of ionic radii is given below (Table 3.2). An ionic radius cannot be determined directly. From X-ray diffraction studies of a crystal like NaF (Na^+ and F^- have the structure $1s^2 2s^2 2p^6$) we can measure the nuclear separation.

TABLE 3.2

Ionic radii of the more common ions (in six-coordination)

Element	z	r/Å	Element	z	r/Å
Ag	+1	1·13	Mg	+2	0·78
Al	+3	0·57	Mn	+2	0·91
Ba	+2	1·43	Mn	+3	0·70
Be	+2	0·34	Mn	+4	0·52
Br	−1	1·96	Na	+1	0·98
Ca	+2	1·06	Ni	+2	0·78
Cd	+2	1·03	O	−2	1·32
Cl	−1	1·81	Rb	+1	1·49
Co	+2	0·82	S	−2	1·74
Cr	+3	0·64	Sc	+3	0·83
Cs	+1	1·65	Si	+4	0·39
Cu	+1	0·96	Sn	+4	0·74
Cu	+2	0·70	Sr	+2	1·27
Eu	+2	1·25	Ti	+3	0·69
F	−1	1·33	Ti	+4	0·64
Fe	+2	0·82	Tl	+1	1·49
Fe	+3	0·67	V	+3	0·65
Hg	+2	1·12	Y	+3	1·06
K	+1	1·33	Zn	+2	0·83
La	+3	1·12	Zr	+4	0·87
Li	+1	0·78			

As the ions are isoelectronic, we would expect Na^+ (nuclear charge 11) to be smaller than F^- (nuclear charge 9). In fact the distance is divided using knowledge for the atomic wave-functions and electron distribution about the two nucleii. Once one radius is defined, we can then assign radii to all ions. It is found that for ionic compounds these radii allow the estimation of distances with an accuracy of a few per cent ($\pm 0{\cdot}1$Å).

Having established size, we can proceed to predict crystal geometry. The number of ions that can pack together at an equilibrium distance will depend on the radius ratio of the ions (Table 3.3). A few simple structures illustrate the way it works.

In NaF the crystal is considered to be made from Na^+ ($r = 0{\cdot}95$) and F^- ($r = 1{\cdot}36$) and from the radius ratio, $r_{Na}/r_F = 0{\cdot}7$ we would predict a co-ordination number of 6. The simple structure of Fig. 3.2 results; a very common structure in AB compounds called the NaCl or rock-salt structure. With CsCl the Cs^+ ion is large enough to pack 8 chloride ions at the equilibrium separation and the structure shown in Fig. 3.3. is formed.

In the mineral rutile (TiO_2), Ti^{4+} ions (radius 0·68) will accommodate six oxygen ions ($r = 1{\cdot}40$) and to comply with neutrality, a 6:3 coordinated structure results (Fig. 3.4).

Potassium chloride is a particularly interesting example. The radius ratio of K^+ and Cl^- is 0·734, very close to the cross-over between the NaF to the CsCl structures. If we keep the internuclear distance constant it can be easily shown that the molar volume of KCl in the CsCl structure is smaller than in the NaF structure. Again, from the lattice-energy equations, we might expect that the ΔH of transition would be similar to the lattice-energy differences and if n stays about constant in the two structures, this will simply reflect the differences in Madelung constants for the two structures. Such simple calculations

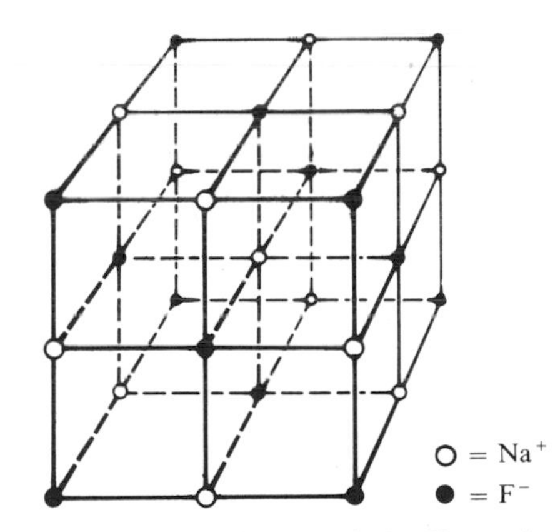

FIG. 3.2. Atomic arrangement in NaF; the NaCl or rock-salt structure.

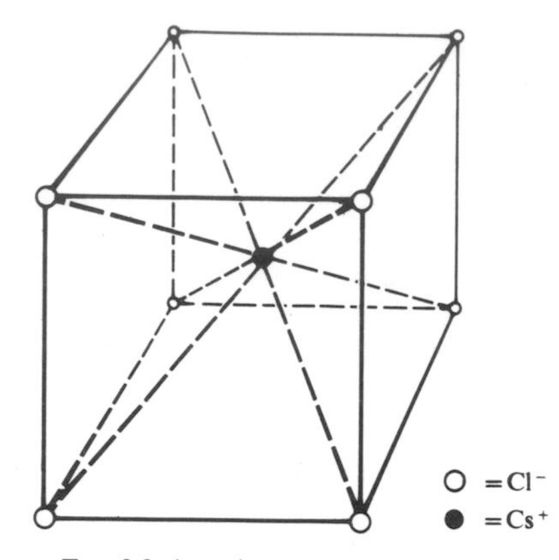

FIG. 3.3. Atomic arrangement in CsCl.

indicate that KCl may show a high-pressure transition from the NaF to the CsCl structures, and the lattice energies and the heat of formation of normal KCl (in the NaF structure) allow us to make a reasonable prediction of where this will occur. In this way it is easy to show that most common minerals will undergo substantial phase changes in the deep earth. The general rule is that high pressures favour large coordination numbers and high temperatures smaller coordination numbers.

The ideal ionic model of a crystal and the prediction of energy–volume relations is extremely valuable in many solid earth systems. But a note of caution must be sounded. Even in simple cases the model will break down and

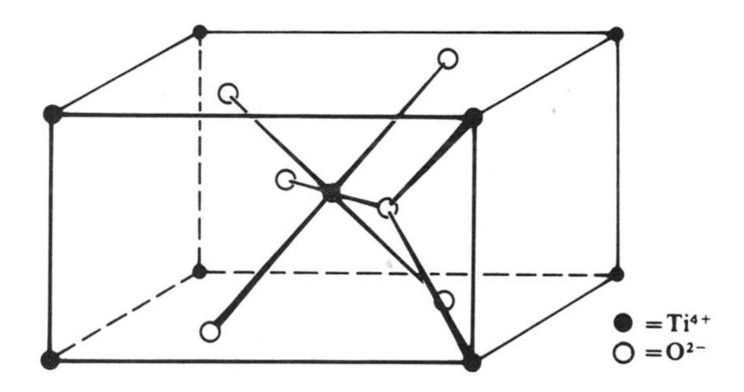

FIG. 3.4. Arrangement of atoms in the rutile structure. The dense modification of SiO_2, stishovite, has the same structure. It is interesting to note that while the molar volume of stishovite is much smaller than for quartz, the Si–O bond-lengths are larger. The total atomic packing determines the density.

TABLE 3.3

Radius ratios and coordination numbers

R_A/R_B	Coordination number of A	Configuration	Example
0–0·155	2	Linear	HF_2^-
0·155–0·225	3	Trigonal planar	CO_3^{2-}
0·225–0·414	4	Tetrahedral	SiO_2
0·414–0·732	4	Square planar	$Ni(CN)_4^{2-}$
	6	Octahedral	NaCl
0·732–1·0	8	Square bipyramid	CsCl

particularly so when transition metals are involved. With the latter, crystal or ligand-field effects resulting from the non-spherical symmetry of the d-orbitals often dominate geochemical behaviour just as they are so important in the inorganic chemistry of these metals. Variations in the radii and energies of their compounds are not explained by a spherical ion model. Silicate minerals dominate the crust and as the Si–O bond must be quite covalent, one must expect that the simple ionic model will have limitations.

Silicate structures

Silicate minerals make up the bulk of the crust and mantle. Attempts to explain silicate chemistry were in a state of total chaos and in a sense were a magnificent exercise in chemical science fiction, until crystal-structure analysis by X-ray methods was introduced. Mineralogists had based classification on crystal form and optical properties rather than on chemistry, an approach later validated.

On an ionic model, one might expect that on account of the radius-ratio of Si^{4+} and O^{2-}, the silica minerals of formula SiO_2 would be built from a tetrahedrally-coordinated silicon atom. On a covalent model one would expect analogies with the carbon atom and a possible sp^3 tetrahedral silicon but with the possibility of a d-electron contribution from empty d-orbitals in the valence shell. The covalent approach is certainly much nearer the truth.

As mentioned before, there are numerous SiO_2 polymorphs. The high-temperature modification β-cristobalite does have a simple structure; silicon atoms occupy a diamond-type lattice with an oxygen bridge. The other important polymorphs, quartz, tridymite, and coesite have more complex structures but are all based on a tetrahedrally-coordinated silicon atom. High-pressure synthesis has revealed an additional polymorph stishovite which exists in the rutile structure, and the compound has a much higher density. Again its stability can be predicted from lattice-energy considerations, and the structure reflects the tendency of silicon to use d-orbitals in octahedral complexes. If free silica exists in the deep mantle of the earth, it will certainly be in this form.

Complex silicates are also based on the tetrahedral (SiO_4) group but in a rather simple manner. In considering aluminosilicates it is always necessary to distinguish octahedrally coordinated AlO_6 groups and AlO_4 groups where Al replaces Si in the structure.

The simplest class of silicates contains isolated SiO_4 groups where no oxygen of the structure forms a bridge between two silicons. Common minerals include:

Mg_2SiO_4	forsterite	olivine
Fe_2SiO_4	fayalite	
Zn_2SiO_4	willemite	
$X_3Y_2Si_3O_{12}$	garnet	

Garnets have a complex chemistry because of extensive solid solutions but formulae like:

$$(Mg,Fe,Mn)_3Al_2Si_3O_{12}$$

or

$$Ca_3(Al, Cr, Fe)_2Si_3O_{12}$$

describe most substitution. There are some rather remarkable garnets where four hydroxy groups $(OH)_4^{4-}$, replace SiO_4^{4-}. The oxygen structure of both groups is similar in dimensions. Thus one may have a continuous series of isostructural garnets:

$$Ca_3Al_2(SiO_4)_3 \quad \text{to} \quad Ca_3Al_2(OH)_{12}.$$

It is possible to pass from the isolated SiO_4 groups to a dimer, Si_2O_7. Thus the common mineral epidote, $Ca_2(Al,Fe)Al_2SiO_4Si_2O_7(O, OH)$, contains both isolated and double units.

After this dimerization, either small rings or infinite chains and bands are formed. Thus beryl $Be_3Al_2Si_6O_{18}$ contains a six-membered ring. Three-membered $(Si_3O_9)^{6-}$ and four-membered $(Si_4O_{12})^{8-}$ rings are also known. The rings in beryl (the mineral shows hexagonal symmetry) are stacked in such a way that a cavity between the rings forms a cage for large molecules; one of nature's own clathrate compounds.

Silicates with an infinite chain structure (Fig. 3.5) occur with the very important mineral family of the pyroxenes. These minerals are significant in many rocks of the crust and upper mantle. The chemistry is very complex owing to an almost universal acceptance of transition metals. Simple members are:

$MgSiO_3$	enstatite
$CaMgSi_2O_6$	diopside
$NaAlSi_2O_6$	jadeite
$NaFeSi_2O_6$	acmite.

In these minerals, some Al may enter the silicon–oxygen chain.

The single chain may become welded into a double chain of unit formula $(Si_4O_{11})^{6-}$ which is essential to the hydrated amphibole family. The chemistry is analogous to the pyroxenes:

$Mg_7Si_8O_{22}(OH)_2$	anthophyllite
$Ca_2Mg_5Si_8O_{22}(OH)_2$	actinolite
$Ca_2Fe_5Si_8O_{22}(OH)_2$	tremolite.

These hydrated phases show extreme thermal stability, the vapour pressure of some being quite low near 1000°C. The hydroxyl groups are sometimes partially replaced by the fluoride ion, an effect common with many hydroxy-silicates. The double chain is strong, and crystals are frequently elongated. Synthetic amphiboles may have a length–diameter ratio near 100.

The metallic ions in amphiboles occupy four distinct structural sites with slightly different coordination numbers or bond angles. When transition metals (Mn^{2+}, Fe^{2+}, Fe^{3+}, Ti^{3+}, Ni^{2+}, etc.) compete for these sites along with Mg^{2+}, Ca^{2+}, Na^+, Al^{3+}, the factors controlling the site distribution are complex. Studies of optical absorption spectra and Mössbauer spectra have clarified this problem and, as is frequently the case, it has been found that ligand-field theory is necessary to explain the competition or site preference.

The large site array available to ions in the amphibole lattice makes them veritable chemical soaks—they will take a little of what is around. In this sense, perhaps only silicate liquids can offer more. To understand the thermodynamics of mixing in such materials is fairly close to a chemical nightmare!

The next stage in polymerization is the formation of an infinite two-dimensional polymer or sheet with unit formula $(Si_4O_{10})^{4-}$. Again a large number of important mineral types fall into this family.

$KAl_2(AlSi_3O_{10})(OH)_2$	muscovite
$KMg_2(AlSi_3O_{10})(OH,F)_2$	phlogopite
$Al_2(Si_4O_{10})(OH)_2$	pyrophyllite
$Al_4(Si_4O_{10})(OH)_8$	kaolinite
$Mg_3(Si_4O_{10})(OH)_2$	talc
$Al(MgFe)_5(AlSi_3O_{10})(OH)_8$	chlorite
$Mg_6(Si_4O_{10})(OH)_8$	serpentine.

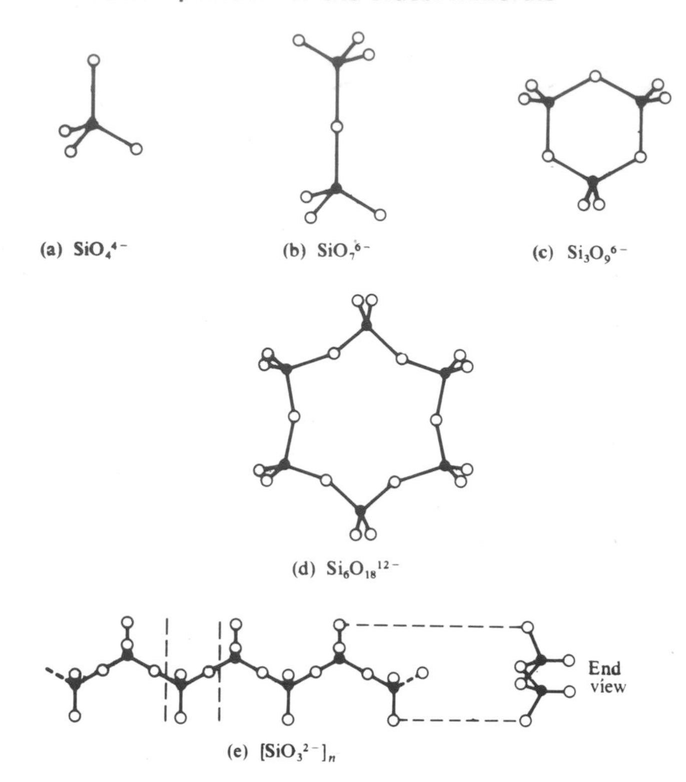

FIG. 3.5. The most common basic units in silicate minerals.

Important mineral groups include the micas, clay minerals, chlorites, and serpentines. All form layer or sandwich structures some of which are illustrated below (Fig. 3.6). The planar morphology of these materials reflects the internal structures. In the case of some clay and serpentine polymorphs, the sheets roll up into tubes. The enormous hydroxyl surfaces of some of these phases explains their peculiar absorbing capacities which at times may be useful to the industrial chemist or a plague to the building engineer. It is interesting to note that the mineral phlogopite is the most thermally-stable simple hydrated substance known; it can be synthesized at 1400°C at moderate water pressures.

Finally we come to the volumetrically most important minerals of the crust, the three-dimensional polymers or framework structures. We have already noted that such structures exist with SiO_2 itself, but if we replace Si by Al, the

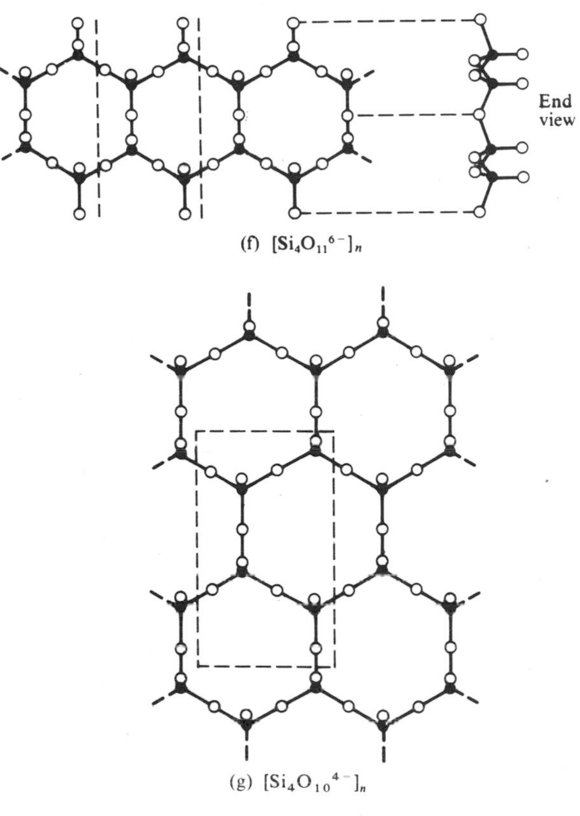

(f) $[Si_4O_{11}^{6-}]_n$

(g) $[Si_4O_{10}^{4-}]_n$

FIG. 3.5. (*cont.*)

neutral SiO_2 group becomes the AlO_2^- group so that more complex aluminosilicates are possible. Thus we find the minerals:

Formula	Mineral	Group
SiO_2	quartz and its polymorphs	
$KAlSi_3O_8$	orthoclase	feldspars
$NaAlSi_3O_8$	albite	feldspars
$CaAl_2Si_2O_8$	anorthite	feldspars
$KFeSi_3O_8$	iron feldspar	
$BaAl_2Si_2O_8$		
$KAlSi_2O_6$	leucite	
$NaAlSiO_4$	nepheline	
$KAlSiO_4$	kalsilite	
$NaAlSi_2O_6 \cdot H_2O$	analcime	zeolites.
$CaAl_2Si_4O_{12} \cdot 4H_2O$	laumontite	zeolites.

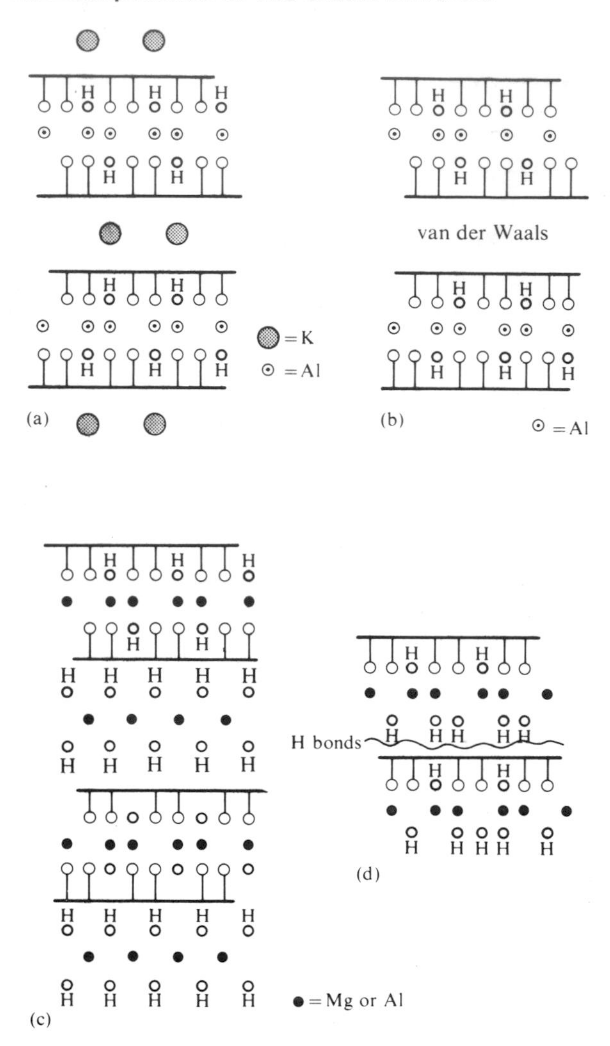

FIG. 3. 6. A schematic representation of the arrangement of the Si–O sheets in sheet-structure silicates. The heavy line represents the layer of Si atoms and an oxygen attached to each projects above or below the sheet. (a) Muscovite, $KAl_2(AlSi_3O_{10})(OH)_2$, double sandwiches of formula $[Al_2(AlSi_3O_{10})(OH)_2]^-$ are linked by potassium ions. (b) Pyrophyllite, $Al_2(Si_4O_{10})(OH)_2$, a double sandwich with no residual charge. (c) Chlorite, $Al(MgFe)_5(AlSi_3O_{10})(OH)_8$. The structure is like the pyrophyllite sandwich with a $Mg(OH)_2$ layer placed between each. (d) Kaolinite, $Al_4Si_4O_{10}(OH)_8$. In this case only a single silicate sheet is present followed by a cation layer and an hydroxy layer.

The feldspars which comprise 50 per cent or so of most common crustal rocks show complex solid solutions and tend to exsolve at low temperatures (p. 66). The zeolites have extremely open low-density frameworks with large channels in the structure. They exhibit facile ion-exchange and molecular-sieve action. Both the water molecules and the cations can often be exchanged for other species of appropriate dimensions. Thus $NaAlSi_2O_6{\cdot}H_2O$ can be readily transformed into $Ca_{\frac{1}{2}}AlSi_2O_6{\cdot}H_2O$ or $AgAlSi_2O_6{\cdot}H_2O$ or even an ammoniated or ethanolated species. They have obvious industrial applications.

In these few notes we have introduced some of the minerals which make up crust and upper mantle. The deep mantle is in a sense a 'no-man's-land' for we know that few of the minerals discussed above will remain unchanged at the great pressures at depth. Add to those we have discussed a few common carbonates ($CaCO_3$, $CaMg(CO_3)_2$, oxides (MnO_2, Fe_2O_3, Fe_3O_4, Al_2O_3), and sulphides (FeS_2, ZnS, PbS, $CuFeS_2$) and we have covered 99 per cent of crustal compounds but only a fraction of a percent of known minerals. The same minerals appear on the moon and in meteorites. In their structures they contain the minor elements and when they undergo changes of state or chemical reactions these elements are redistributed. We cannot begin to understand detailed geochemistry without some knowledge of these complex compounds.

4. Rocks

THE branch of science concerned with the nature and origin of rocks is called petrology. Rocks are heterogeneous objects made from an assemblage of mineral phases. The number of minerals in a given rock rarely exceeds ten or so and sometimes is much less. It should be noted that organisms synthesize minerals in their cells and such debris often make a massive contribution to crustal rocks.

Rock classification and nomenclature is complex and may be based on some dominant feature of origin, chemistry, texture or fabric, and the like. Here we shall be mainly concerned with the three major types:

Igneous rocks—derived from the solidification of silicate liquids.

Sedimentary rocks—formed on the surface by transport and chemical processes.

Metamorphic rocks—formed by the recrystallization of either igneous or sedimentary rocks due to changes in the *P–T* regime or the position of the rocks in the earth.

The major processes of formation involve liquids: with igneous rocks a high-temperature silicate melt; with sedimentary rocks surface aqueous fluids of the oceans, rivers, etc.; metamorphic processes normally involve high-temperature aqueous fluids.

Igneous rocks, or their metamorphic equivalents, form the bulk of the earth's crust. Sediments represent a relatively thin veneer, seldom more than a few km thick. There is good evidence that igneous phenomena and volcanism were at least as intense in the past and possibly more intense. There is evidence too, that in the past liquids reached the surface at higher temperatures, a fact possibly to be correlated with a larger degree of melting throughout the young earth. It would be surprising if this was not so, for radioactive heating must have been more intense in the past.

Igneous rocks

Volcanism is one of the most spectacular phenomena proving that the interior of the earth is hot. During volcanic events silicate liquids appear at the surface with temperatures in the range 800–1300°C. Such liquids may cool rapidly to form partially glassy rocks, or cool more slowly to form crystalline mixtures. Volcanic events are spread through space and time, but at the present such phenomena are concentrated at the ocean ridges and the highly seismic regions of some continental margins (e.g. the west of the Americas). If a melt is injected at depth it will cool more slowly and will produce larger average crystal sizes. Igneous rocks formed at depth are termed plutonic.*

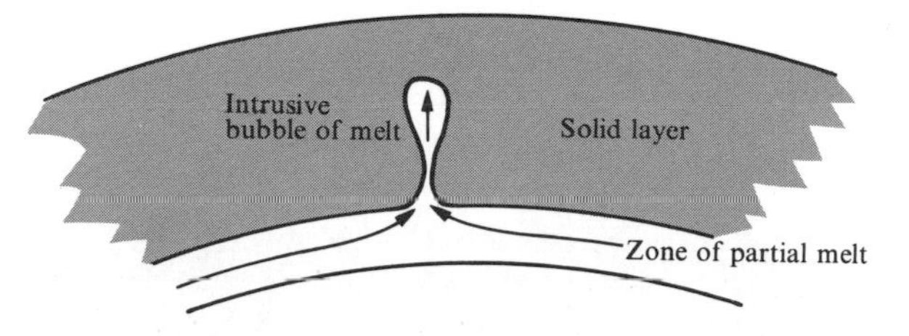

FIG. 4.1. General behaviour of a gravitationally-unstable layer. A low-density melt forms bulges or bubbles which invade the stiffer overburden. At high levels cracks may form and surface extrusion can occur.

At any point in our largely solid earth, heat production is such that, if melting commences, the motion of the melt will remove heat more efficiently than any conductive or radiative heat-transport process. Thus melting acts as the thermal buffer at any point in the earth.

Imagine a uniform radial shell at depth which commences to melt. Silicate liquids are less dense than solids so that a gravitational instability is generated (the classic Taylor instability of fluid dynamics). Simple models show that the unstable layer will develop bulges and eventually plumes* (bubbles) of the light material (Fig. 4.1, 4.2) will rise. The mathematics of this situation is rather well known and one can predict the order of magnitude of the separation between such rising plumes on the basis of knowledge of densities, thicknesses, and viscosities of the materials. It is rather remarkable that volcanic events on earth may be rationalized by such an approach.

At the present time, melting occurs in two major situations in the earth and produces three major types of chemical composition. High thermal gradients occur near the major ocean ridges and are associated with magmatism.* We could consider these sites (Fig. 4.3) as those where convection currents rise. The melts are formed at depths of 100–200 km, and melting is associated with the low-velocity zone (p. 6). This liquid represents the product of partial fusion of the upper mantle. The volcanic liquid so produced is called basalt and an average chemical composition is shown in Table 4.1.

This basalt and its intrusive equivalent gabbro,* forms the new ocean-floor material. The material spreads out from the ocean ridge, and this spreading rate is measurable from paleomagnetic* studies of the magnetization of basalt and a knowledge of the timing of magnetic reversals of the earth's field. Basalt contains the mineral magnetite which carries most of the magnetic record. Typical spreading rates are 1–10 cm per year.

Near ocean trenches, the circle is closed. Ocean floor crust descends into the upper mantle. It carries with it pieces of the continent, ocean-floor sediments, etc. As it descends it heats up and eventually partially melts. The melting of this heterogeneous material produces the igneous rock termed andesite

FIG. 4.2. An air photograph showing more-or-less circular granite bubbles rising into the crust of Saudi Arabia. The large circular objects are about 10 km in diameter. (Photo by courtesy of the Directorate of Mineral Resources, Saudi Arabia.)

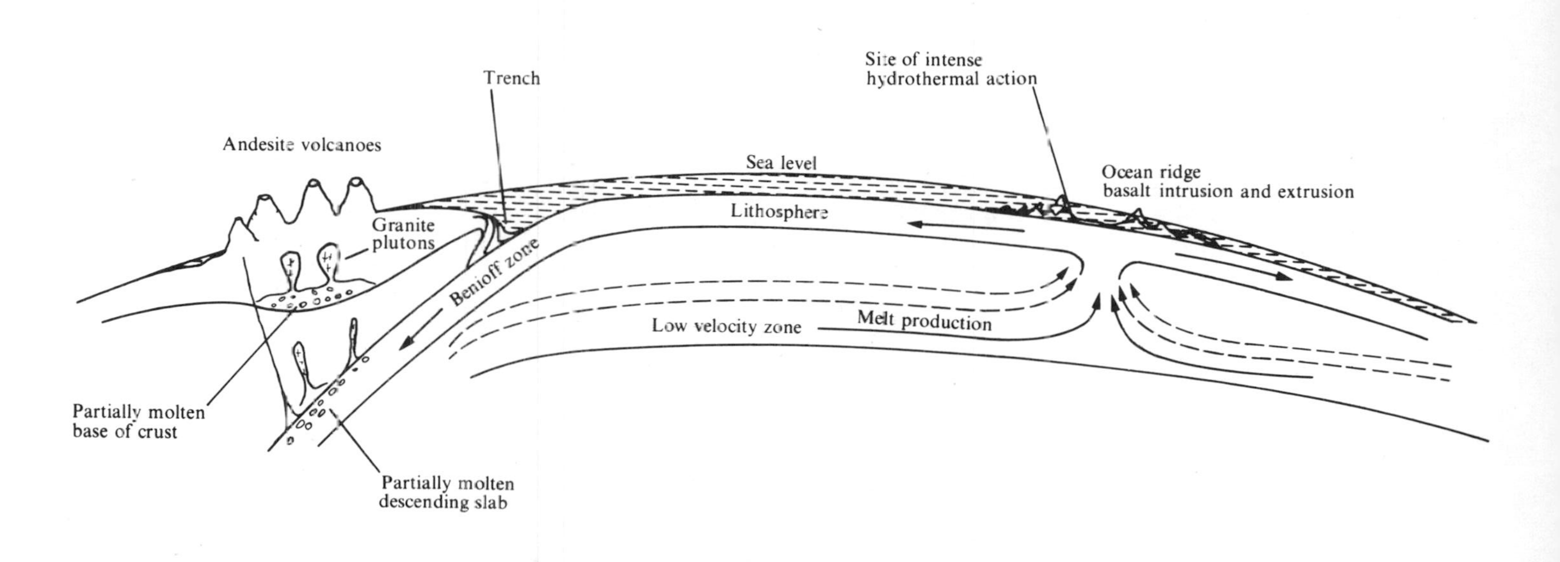

FIG. 4.3. Schematic diagram of a modern convection cell. Partial melting occurs in the low-velocity zone and basalt melt is intruded at the ocean ridge. This is a zone of high heat flow. The ocean floor flows away from the ridge (velocity 1–10 cm y^{-1}) and may descend at a continental margin as shown. Where descent commences an ocean trench* forms. The trench region is one of low heat flow, for cold material descends into the hotter mantle. The ocean floor crust and part of the sediment load returns to the mantle. At depth it may melt producing andesitic liquids which form surface volcanoes. The region is one of high heat flow. The heat flux may lead to melting of the base of the crust and granite plutons* rise beneath the andesitic lava flows. A typical area illustrating the model could be a strip from the East Pacific rise to the Peru trench to the Andes mountain chain; about latitude 30° South.

TABLE 4.1
Average composition (per cent) of main magma types; a range is shown for granitic types

Component	Basalt	Andesite	Granite
SiO_2	49·74	59·57	65·01–70·18
TiO_2	1·36	0·77	0·57– 0·39
Al_2O_3	16·57	17·30	15·94–14·47
Fe_2O_3	2·33	3·33	1·74– 1·57
FeO	6·95	3·13	2·65– 1·78
MnO	0·17	0·18	0·07– 0·12
MgO	7·54	2·75	1·91– 0·88
CaO	11·50	5·79	4·42– 1·99
Na_2O	2·79	3·58	3·70– 3·48
K_2O	0·19	2·04	2·75– 4·11
H_2O	0·73	1·26	1·04– 0·84
P_2O_5	0·13	0·26	0·20– 0·19

(Table 4.1). Such rocks are formed in vast volumes in mountain ranges associated with ocean-trench systems (e.g. the west of the Americas—andesites make vast contributions to the Andes and most modern mountain ranges). The geometry of the cycle is shown in Fig. 4.3. The descending slab is associated with complex seismic activity and most of the modern world's destructive earthquakes are associated with this phase of motion. The region is termed a Benioff zone. As the andesite liquids rise from the descending slab, the thermal gradient in the crust is elevated and eventually the light crust may commence to fuse; the fusion product we will term granite. These liquids form the core of many mountain ranges or produce great gas-rich volcanic events on the continental surface. Because granitic liquids have high viscosities ($\eta \sim 10^{10}$ poises), the liquids often quench to glasses if erupted on the surface.

Thus igneous events are concentrated in regions of rising and descending convection currents. But the earth also appears to be a little thermally 'noisy' and occasionally volcanic events occur away from these dominant sites of activity.

The range of composition of melts formed at these regions of activity is very large. The reasons are obvious; melting occurs over a range of P–T conditions in a multicomponent and perhaps heterogeneous system; the melt is generated at depth and may partially crystallize or dissolve its wall rocks during ascent.

We shall return to the details of fusion and crystallization processes in Chapter 6. Suffice to note here that *via* igneous processes the surface material is continuously recycled.

Sedimentary rocks

In a sense, the primary materials of the earth's surface are volcanic rocks. But at all places on the surface, air, water, and organisms and their debris attack rocks and form sediments. It is obvious that igneous rocks which are generated in a high P–T, rather reducing, and rather dry environment, will hardly be in equilibrium with the surface environment. They are rapidly eroded and corroded.

Some sediments result simply from the mechanical break up of rocks (for example along grain boundaries) and the transport of the debris. We could term these mechanical sediments. But in general, chemical attack is universal and the processes which loosen crystals or generate weakness in rocks may involve chemical processes.

The dominant chemical processes occurring with surface rocks are as follows.

Solution. For example, quartz has a solubility of about 10 kg per 10^6 kg of water at normal conditions. The solution reaction is probably:

$$SiO_2 + 2H_2O \longrightarrow Si(OH)_4(aq).$$

This solubility (2×10^{-4} mol dm^{-3}) seems small but it has been estimated that, in general, the land surface is lowered by about 1 cm in 1000 years by solution alone.

Minerals like SiO_2, NaCl dissolve congruently in water. Most minerals dissolve incongruently. For example the common feldspar mineral orthoclase may dissolve by a mechanism such as:

$$2KAlSi_3O_8 + 11H_2O \longrightarrow \underset{\text{kaolinite}}{Al_2Si_2O_5(OH)_4} + 4Si(OH)_4(aq) + 2K^+(aq) + 2OH^-(aq)^-.$$

A new solid is formed (a clay mineral) while silica and potassium are leached and transported away. The solubility depends on pH and may be drastically influenced by biological activity. Such reactions are involved in soil formation and in tropical countries with intense biological activity, the end product of the leaching of most rocks is a soil dominated by SiO_2–Al_2O_3–Fe_2O_3; a soil just about useless for food production.

Oxidation processes are also important in the weathering zone. Thus the common mineral olivine of basaltic rocks, or rather its component Fe_2SiO_4, may weather according to a reaction:

$$Fe_2SiO_4 + \tfrac{1}{2}O_2 + 3H_2O \longrightarrow \underset{\text{geothite}}{2FeO(OH)} + Si(OH)_4(aq).$$

Most igneous minerals contain iron in the ferrous state while iron in most sediments is in the ferric state. Other transition elements like Mn, Co, are affected in the same way.

Hydration reactions occur with almost all igneous minerals; feldspars are converted to clays and zeolites; pyroxenes, amphiboles, and micas are converted to chlorites. The average igneous rock contains less than 1 per cent water while sediments contain 5–10 per cent chemically bound water. In addition,

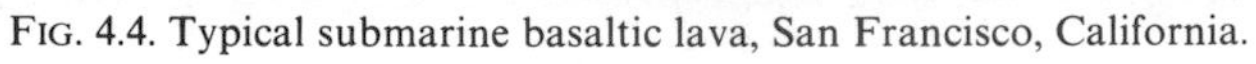

FIG. 4.4. Typical submarine basaltic lava, San Francisco, California.

FIG. 4.5. Deposits of silica and iron manganese oxides (rich in copper, nickel) which rest on the submarine lavas. It is very probable that these materials were stripped from the lavas by heated circulating sea water. The silica is deposited biologically.

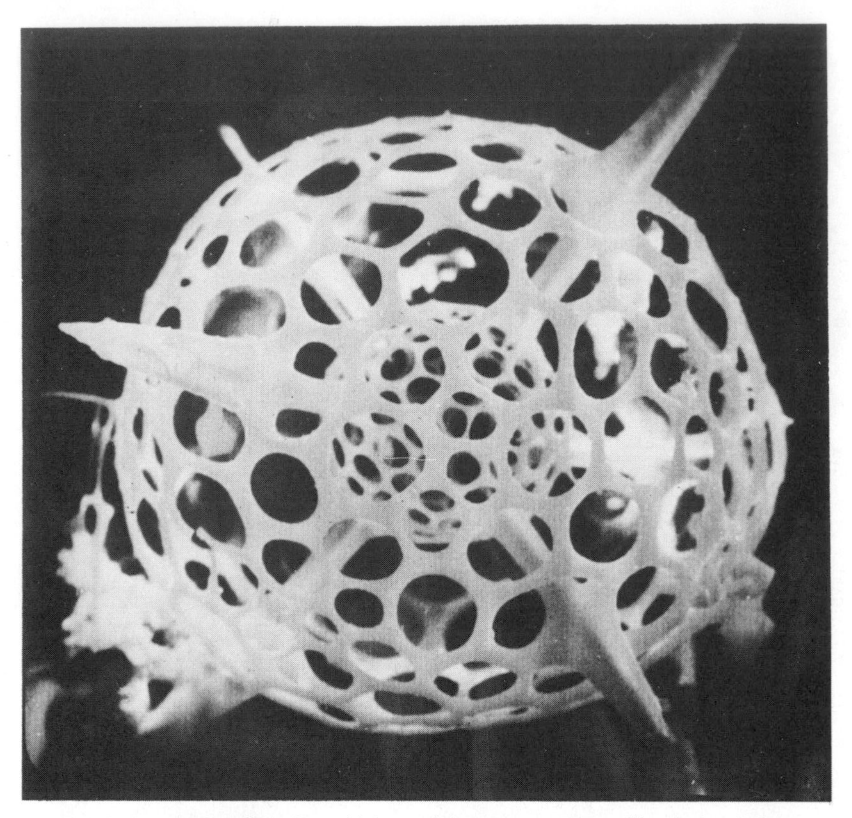

FIG. 4.6. A typical radiolarian which precipitates amorphous silica in its skeleton. Scanning electron microscope picture (× 1000).

some organisms fix CO_2 in the atmosphere by precipitating carbonates.

As well as these complex, generally sluggish, heterogeneous processes which occur universally in the surface environment, important biochemical mineral precipitates are formed. Many marine organisms secrete calcium carbonate in their skeletons; radiolaria precipitate amorphous silica, Fig. 4.4, 4.5, 4.6; chitons grow teeth of magnetite; man of apatite. The array is formidable. Bacteria may be responsible for massive precipitation of sulphides and iron oxides. In favourable cases, thousands of km^3 of rock are dominated by such debris as in fossil coral reef formations. Such biological activity has an important influence in partially controlling the CO_2–O_2 balance of the atmosphere (see p. 84). Biological activity is also responsible for introducing complex organic materials into sediments. Accumulation and alteration may produce deposits of coal, oil etc. (p. 99).

TABLE 4.2
Some major types of sedimentary rocks

Rock type	Mineralogy	Chemistry
Sandstone	quartz, feldspars	dominated by SiO_2 (+K, Na, Ca, Al)
Shale	clay minerals, chlorites, carbonates	Al_2O_3–SiO_2–H_2O
Limestone	calcite, dolomite	$CaCO_3$, $MgCO_3$
Chert	quartz, hematite	SiO_2 (+ minor Fe_2O_3, MnO_2)
Phosphorite	apatite	calcium phosphate
Soil	complex clay minerals, quartz, organic materials	Al_2O_3–SiO_2–H_2O
Laterite	bauxite, hematite	Al_2O_3–Fe_2O_3–SiO_2
Evaporite	halite, gypsum	NaCl, $CaSO_4$, $CaCO_3$ (sometimes borates)

Finally, simple precipitation reactions occur in the aqueous environment. The most obvious cases involve the total evaporation of surface waters and the formation of evaporites. Products such as $CaCO_3$, $CaSO_4$, NaCl, $MgSO_4$, KCl, and complex borates result. But in the oceans generally, precipitation of complex clay minerals, zeolites, carbonates, sulphates, tends to maintain a constant balance of addition and subtraction. We shall look in more detail at examples of this balance in Chapter 8.

It is rather hopeless to give chemical compositions of sedimentary rocks; the degree of variation is extreme. But some general features of common types are summarized in Table 4.2.

Metamorphic rocks

After primary crystallization, sedimentary and igneous rocks may be subjected to new physical conditions imposed by tectonic forces. Thus, sediments may be buried to depths of 30 km where the regional pressure is 10 000 atmospheres and T about 600°C. New minerals form in response to the changing conditions and the rock with its new mineral assemblage is termed metamorphic. Because any sediment or igneous rock may be subjected to a very wide range of physical conditions, the number of metamorphic minerals is large.

Metamorphic processes affect solid materials and the forces which lead to burial and uplift imply the operation of stress. This stress often results in preferred orientation of the new mineral grains and a rather well developed fabric is one of the general characteristics of many metamorphic rocks. This preferred orientation is termed schistosity and the rocks schists. When clay sediments are metamorphosed the newly-formed mica crystals of the metamorphic rock, slate, show an almost perfect planar fabric.

Thus metamorphism results from progressive heating (or cooling) and compression (or decompression) of essentially solid materials. Normally, moderate stresses (perhaps a few hundred bars) operate. While a metamorphic rock is solid throughout its history, gases like water and carbon dioxide are eliminated or absorbed and are themselves catalysts for the reactions. Some, but few, processes occur in the solid state. When water is lost on a large scale as it is during burial, elements are carried upwards by the escaping fluids and may be deposited during cooling and pressure release. Many ore deposits are formed by such processes (Cu, Ag, Au, W, Zn, etc.).

Clearly, as most of the earth is solid, and as the earth is a dynamic system, metamorphic rocks dominate both the observed and hidden parts of the earth. Much of our mineral wealth comes from metamorphic rocks. Because such rocks, or the minerals within them, record changes in P and T, these rocks record the dynamic history of the earth (Chapter 6).

5. Chemical compositions of objects in space and the whole earth

WE shall now turn to estimates of the chemical composition of the whole earth. It is a formidable task and certainly present estimates will prove to be imperfect. As we have noted, the accessible part of the earth makes up less than 1 per cent of the mass and for many reasons can hardly represent the interior. As man probes other planets in the solar system and the sample from which we derive our ideas about accumulation becomes enlarged, refinements will almost certainly follow.

Any model of total-earth composition must fit a number of constraints:

(1) The composition must be reasonable in relation to the compositions of other objects in the solar system and the universe in general. It is unlikely that the elements in our earth were synthesized by some unique nuclear event.

(2) The composition must account for the data of geophysics found for each part of a layered earth; seismic wave velocities, magnetic field, heat production, and the like.

(3) The composition of the deeper layers must be appropriate to explain the composition of melts coming from depth as a product of partial fusion.

(4) The distribution of elements in the earth should be reasonable in terms of the thermodynamic principles governing matter in a gravitational field.

It turns out that all these constraints make possible choices of materials quite limited and, if our knowledge of matter at very high pressure were better, the choice might be quite small.

Extraterrestrial abundances

Let us first look briefly at the composition of other materials which exist in space. For our solar system, the sun makes up over 99·8 per cent of the mass. All theories of origin suggest that the planets are accumulates from an early phase of development of the entire system so that the sun's composition should be reflected in the composition of individual planets. One might also expect that the sun might be similar to matter in the Universe in general but stars evolve and their chemistry must evolve with them.

From studies of the solar spectrum the relative abundances of about seventy elements have been determined. Other elements must also be present but for one reason or another, the spectrum cannot be observed or resolved on earth. Observations from space could do better. As with all stars where heat is produced by a fusion process, hydrogen and helium dominate. But, if we examine the data (Table 5.1), we notice that while in a general way, abundance falls off with atomic number there are high spots (e.g. at silicon, iron, etc.). These magic numbers have been known for a long time and can be correlated with nuclear binding energies and with general aspects of nuclear stability.

We shall not discuss the problem deeply here except to note that a complex series of nuclear reactions are needed to rationalize the overall element-distribution pattern. These include reactions between both light and heavy nuclei, a reaction series which evolves throughout the history of a given star. Different types of stars give slightly different ratios but solar abundances are a major source of detailed information. When these are considered it must be stressed that the spectra observed originate from the outer layers of the sun and we may well ask if the sun is completely mixed. We can note at this point, that differences between our crust and the sun are apparent. For example, if we disregard the light volatile elements, the K : Si ratio of the crust is about 1 : 8 while for the sun this ratio is 1 : 700. We might suspect that this heat-producing element potassium is highly concentrated at the earth's surface.

Along with the sun, meteorites have long attracted attention as samples of matter from other parts of the solar system. Many pieces of evidence indicate that some of them represent fragments of a disrupted larger body and hence they may shed some light on what we would see if our planet was disrupted.

Meteorites are of many types and a great number have been examined in immense detail. Certain features are common to all. It has been estimated that several thousand tons of this material arrive on our surface annually and we know that some of the masses have been large enough to generate craters several kilometers in diameter. Normally the surrounding rocks show evidence for intense shock-wave phenomena, including the formation of very high-pressure phases such as stishovite.

Basically, there are three main types of meteorites; those dominated by iron, a mixture of iron and silicates, and silicate meteorites. The silicate meteorites are divided into the chondrites (from the presence of small rounded objects called chondrules*) and the achondrites. Their mineralogy is in part earth-like but many minerals are known only from meteorites. Naturally there is a much greater tendency to *find* iron meteorites because the silicate types when weathered will blend with crustal rocks. So to obtain any realistic picture of meteorite statistics one must rely on recorded 'falls'. Of '*finds*' metals meteorites make up 59 per cent but of observed '*falls*' only 6 per cent, while the remainder are 92 per cent silicate types and 2 per cent metal–silicate mixtures. Thus in using meteorites for planetary data, most attention is given to the silicate types.

Metal meteorites are dominated by iron (normally about 90 per cent) with some nickel (6–9 per cent) and minor amounts of Co, Cu, P, S, and C. The main mineral phases are taeinite, an Fe–Ni alloy; kamacite, nearly pure iron,

TABLE 5.1

Abundance of the elements (*relative to* Si = 10^6)

Element	Cosmic	Sun	Element	Cosmic	Sun
H	$3·2 \times 10^{10}$	$3·2 \times 10^{10}$	Pd	1·0	0·5
He	$2·6 \times 10^9$		Ag	0·26	0·04
Li	38	0·3	Cd	0·89	0·91
Be	7	7·2	In	0·1	0·46
B	6		Sn	1·33	1·10
C	$1·66 \times 10^7$	$1·66 \times 10^7$	Sb	0·15	2·8
N	3×10^6	3×10^6	Te	3·0	
O	$2·9 \times 10^7$	$2·9 \times 10^7$	I	0·46	
F	10^3		Xe	3·15	
Ne	$2·9 \times 10^6$		Cs	0·25	
Na	$4·2 \times 10^4$	$6·3 \times 10^4$	Ba	4·0	4·0
Mg	1×10^6	8×10^5	La	0·38	
Al	9×10^4	5×10^4	Ce	1·0	
Si	10^6	10^6	Pr	0·16	
P	9×10^3	7×10^3	Nd	0·69	
S	6×10^5	6×10^5	Sm	0·24	
Cl	$1·8 \times 10^3$		En	0·08	
Ar	$2·4 \times 10^5$		Gd	0·33	
K	$2·9 \times 10^3$	$1·6 \times 10^3$	Tb	0·05	
Ca	$7·3 \times 10^4$	$4·5 \times 10^4$	Dy	0·33	
Sc	29	21	Ho	0·08	
Ti	$3·1 \times 10^3$	$1·5 \times 10^3$	Er	0·21	
V	590	158	Tm	0·03	
Cr	$1·2 \times 10^4$	5×10^3	Yb	0·18	1·07
Mn	$6·3 \times 10^3$	$2·5 \times 10^3$	Ln	0·03	
Fe	$8·5 \times 10^4$	$1·2 \times 10^5$	Hf	0·16	
Co	750	1380	Ta	0·02	
Ni	$1·5 \times 10^4$	$2·6 \times 10^4$	W	0·11	
Cu	39	$3·5 \times 10^3$	Re	0·05	
Zn	200	800	Os	0·73	
Ga	9	7·2	Ir	0·5	
Ge	134	62	Pt	1·2	
As	4·4		Au	0·13	
Se	19		Hg	0·27	
Br	4		Tl	0·11	
Kr	20		Pb	2·2	2·5
Rb	5	9·5	Bi	0·14	
Sr	21	13	Th	0·07	
Y	3·6	5·6	U	0·04	
Zr	23	54			
Nb	0·8	2·8			
Mo	2·4	2·5			
Ru	1·6	0·9			
Rh	0·26	0·19			

and small amounts of other phases such as iron carbide, nitrides, and phosphides.

The silicate meteorites are dominated by phases such as olivine and pyroxenes. A large range of sulphides occur (of Ni, Cu, Fe, Zn, etc.) but also some unusual sulphides of elements like K and Ca. One class of chondritic meteorites is most intriguing in that they contain large amounts of carbon compounds, water, and even hydrated silicates like chlorite. Analyses of some typical silicate meteorites are given in Table 5.2. The carbonaceous matter in carbon-rich meteorites is complex. It is dominated by paraffinoid hydrocarbons but includes fatty acids, porphyrins, nucleic acids, and amino acids. Certainly complex molecules exist and naturally these are of great interest to those concerned with the origin of life.

There are many views on the origin of these complex and old (mostly about $4{\cdot}5 \times 10^9$ years) materials. Data based on the presence of the daughter elements of short-lived isotopes perhaps indicates that they accumulated very rapidly after nucleosynthesis. The varieties now observed could be explained by a series of metamorphic and igneous events acting on asteroidal-sized bodies which were later disrupted. If this is the case, the carbonaceous chondrites might represent the type of matter that condenses during a planet-forming event. Further, if one contemplates the analyses of Table 5.2 and allows for some loss of volatile elements, it is not too difficult to approach an earth model.

TABLE 5.2
Some meteorite analyses

Component	1	2	3	4
SiO_2	23·08	27·31	54·01	36·52
MgO	15·56	19·00	35·92	23·48
FeO	10·32	20·06	0·97	8·87
Al_2O_3	1·77	2·31	0·67	2·43
CaO	1·51	2·03	0·91	1·82
Na_2O	0·76	0·54	1·32	0·85
K_2O	0·07	0·05	0·10	0·14
Cr_2O_3	0·28	0·39	0·06	0·36
MnO	0·19	0·17	0·14	0·25
TiO_2	0·08	0·10	0·06	0·13
P_2O_5	0·27	0·27	0·22	0·23
H_2O	20·54	13·23	1·14	0·33
NiO	1·17	1·56	0·26	
FeS	16·88	8·58	1·25	5·35
C	3·62	2·44		0·10
Metal	0·13	0·16	2·46	18·90

(1 and 2 are for carbonaceous chondrites; 3 enstatite achondrites; 4 high-Fe chondrites.)

Many workers have approached the problem of cosmic element abundances. If stars evolve and their nuclear furnaces evolve also, it is clear that this is a most difficult task for, in any given stellar system, the pattern will change with time. Certainly most of the universe is hydrogen. Attempts have been made to make intelligent estimates. The basic building blocks of these estimates are data from stellar spectra, meteorites, and theories of nuclear synthesis by fusion processes. The type of values obtained are shown in Table 5.1.

Composition of the earth

Most calculation of the composition of the entire earth make the following assumptions:

(1) The core is like the Fe–Ni alloys of meteorites. But experimental data on seismic wave velocities in the core indicate that the average atomic number is rather less than iron or nickel and that a light element is added. These could be elements like C, N, S, H, etc. Again by analogy with meteorites, it seems reasonable that this component might be dominated by sulphur.

(2) The material of the mantle is assumed to be like the oxidized fraction of silicate meteorites, and is dominated by $MgO–SiO_2$, but contains enough Na–Ca–Al etc. to sweat out a basaltic liquid fraction when heated. Solid inclusions which arrive at the surface in basalt liquids may be samples of such upper mantle and they do in fact have compositions in accord with such a model. Experimental studies have shown that seismic velocity changes in the mantle could be accounted for by such compositions; the discontinuity at 350–450 km being caused by the olivine structure passing to a spinel (oxide) structure and later events to a complete change to mixed high-pressure oxide phases. Very high pressures are leading to a basic structural simplification, the inverse of a high-temperature trend towards mixing.

(3) The crust represents an accumulation of the most fusible components.

Along these lines the earth composition shown in Table 5.3 is derived. While it must be imperfect, it fits most of the facts. It also shows that when the sun and solar system formed, there was a considerable degree of fractionation of

TABLE 5.3

Chemical composition of the earth (after Mason)

Fe	35%	S	1·9%	Co	0·13%
O	30	Ca	1·1	P	0·10
Si	15	Al	1·1	K	0·07
Mg	13	Na	0·57	Ti	0·05
Ni	2·4	Cr	0·26	Mn	0·22

the elements. Perhaps each planet is a little unique and perhaps too much emphasis is placed on meteorites which represent an unsuccessul planet. By the end of this century we will know much more.

Element distribution in the earth

Does the type of earth model indicated above make chemical sense and what happens to all the elements present in minor amounts? Perhaps we could envisage an experiment where we take all the material of the earth, cold, and in a large beaker without a gravitational field. We note that the metallic elements exceed the number of non-metallic elements so that metal must be left over. The first reactions then involve a competition of the type:

$$M^1O + M^2 \rightarrow M^2O + M^1$$

$$M^1S + M^2 \rightarrow M^2S + M^1$$

$$M^1\,(\text{silicate}) + M^2 \rightarrow M^2\,(\text{silicate}) + M^1.$$

Given thermodynamic data we can answer this question at low pressures, but not realistically at core pressures. There can be some surprises. Because of chemical and mineralogical experience, we tend to think that elements like potassium will tend to form oxides while elements like Fe may tend to form sulphides but for the reaction:

$$K_2S + FeO \rightarrow FeS + K_2O,$$

$\Delta G^{\ominus} = +230\,\text{kJ mol}^{-1}$ and sulphur will go with potassium. But if we put potassium in a silicate matrix, the equilibrium goes in the opposite direction. We must be just a little careful in these arguments for the affinity of an element for any other depends on the bulk chemistry of the system and the phases formed. When we allow for such factors, we would expect to see silicates and aluminosilicates of the most electropositive elements, and a metal phase which would carry the heavy inert elements like Pt and Au.

Next in our experiment we must switch on the gravitational field and impose a thermal gradient. We now face a problem where we must solve a series of equilibrium equations for each component such as:

$$d\mu_i = O = \left(\frac{\partial\mu_i}{\partial T}\right)_{P,h,N} dT + \left(\frac{\partial\mu_i}{\partial P}\right)_{T,h,N} dP + \left(\frac{\partial\mu_i}{\partial h}\right)_{T,P,N} dh + \sum_{ij}\left(\frac{\partial\mu_i}{\partial N_j}\right)_{P,\,T,\,h,\,N_k} dN_j$$

an equation which describes the variation in the chemical potential of our chosen element when temperature T, pressure P, height in the gravitational field h and concentration of other elements, are all varied. It is quite beyond us to carry out any rigorous solution to this equation as we do not know species or their molar volumes etc. for the deep earth. But one can make reasonable guesses, by first considering the major elements. This would show that light oxygen and its compounds would be near the surface. As oxygen must enrich upwards, other elements with a large oxygen affinity will also enrich upwards according to their densities. Thus elements like uranium and tungsten will follow oxygen and concentrate in the crust.

Elements like gold and platinum, with noble characters, will sink and alloy with the metal phases. This type of partition has been observed in meteorites and slags. Long ago, geochemists classified elements as siderophile* (metal-loving), chalcophile* (forming sulphides), and lithophile* (forming silicates). The elements which are siderophile and chalcophile will tend to be impoverished in the crust. Thus the solar-system abundance ratio of Au:Si is about 2×10^{-7} while in the crust of the earth it is about 2×10^{-8}. The U:Si ratio of the solar system is about 10^{-8}, while in the crust it is about 10^{-5}.

Problems abound. We do not know how much potassium is in the earth and this is important in terms of heat production. We do not understand the competition reactions listed above at 10^6 atmospheres. But we can say that the earth is not in a state of perfect equilibrium because light materials are still arriving at the surface where gravitationally they are more stable. Either

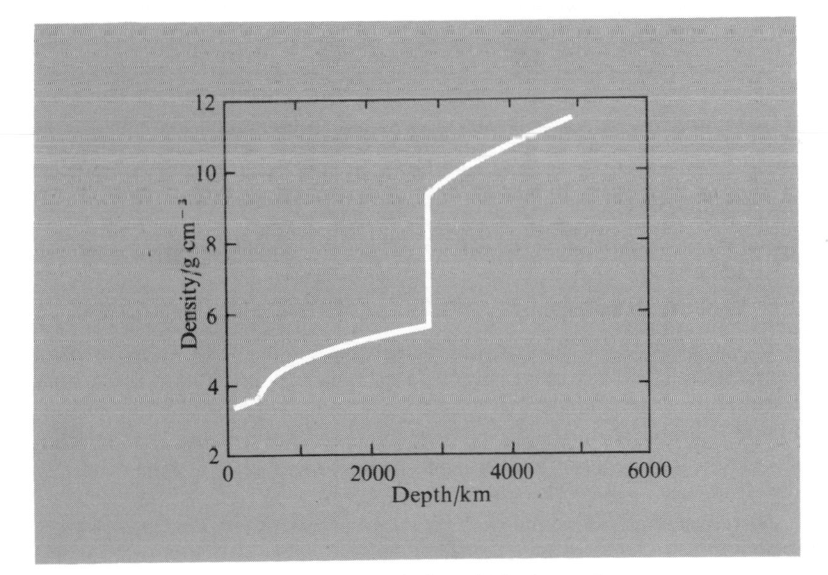

FIG. 5.1. Density variation within the earth.

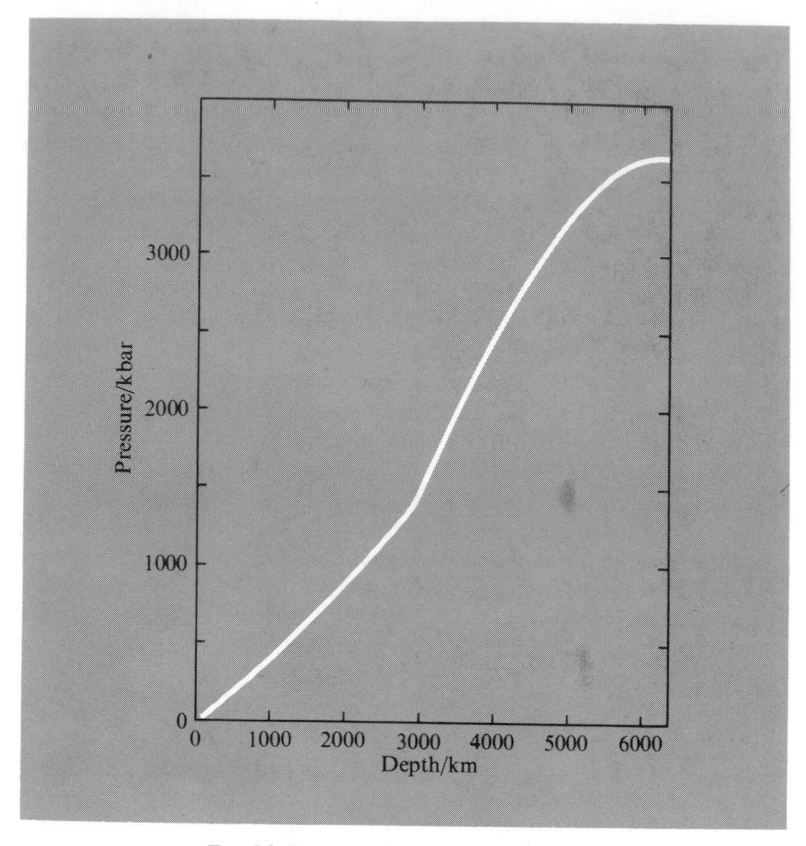

FIG. 5.2. Pressure variation within the earth.

the earth is not yet perfectly differentiated or it is in a steady state of partial mixing.

P–T Variation in the earth

Having arrived at a model of chemical variation for the earth, we can now fit the seismic velocity observations and the mass distribution observations together to obtain a model for *P* variation. Unique solutions are not possible but the type of variations shown in Figs. 5.1 and 5.2 seem plausible. Obviously, if we know the chemistry at depth, and the melting relations of the materials, this places constraints on possible temperatures.

There are some very important general considerations which arise from these numbers. Pressure at the centre of the earth exceeds 3×10^6 atmospheres while *T* is probably around 4000°C. Consider any possible chemical process in solids, A → B. The state of equilibrium of the system depends on $\Delta G = G_B - G_A$ and the variations:

$$\left(\frac{\partial \Delta G}{\partial P}\right)_T = \Delta V \quad \text{and} \quad \left(\frac{\partial \Delta G}{\partial T}\right)_P = -\Delta S.$$

For most processes in solids, ΔS and ΔV have the same sign. If volume decreases, the entropy decreases, as we would expect if entropy is related to uncertainty or chaos.

Averaging over many solid–solid transformations we find that for a ΔV of 1 $cm^3\ mol^{-1}$, ΔS will be of the order of 0·5 cal mol^{-1} deg (2·1 J $K^{-1}\ mol^{-1}$). Thus for a process with these values of ΔS and ΔV, the variation in ΔG is going from crust to inner core will be: for temperature, $\Delta S\,\Delta T = 0{\cdot}5 \times 4000 = 2000$ cals mol^{-1} (8400 J mol^{-1}), and for pressure, $\Delta V\,\Delta P = 1 \times 3 \times 10^6$ cm^3 atms

$$\sim 70\,000 \text{ cals mol}^{-1}\ (294\,000 \text{ J mol}^{-1}).†$$

Pressure is much more significant than temperature for the entire earth. The $\Delta V\ \Delta P$ term is so large that, even for small volume changes, the energy gained by contraction is comparable to the energy of chemical bonds. It is not a surprise that familiar surface materials change structure at depth. On the other hand in the crust, where heat production due to U, K, and Th decay is larger and the thermal gradient is steeper, T and P variation have roughly equivalent thermodynamic influence.

† Note that $\Delta T = T_{core} - T_{crust}$ and $\Delta P = P_{core} - P_{crust}$.

6. Mineral reactions—phase changes

WE observe only a trivial part of the earth's mass but yet we wish to construct a complete history. If this is to be achieved it is obvious that we must extract the maximum possible amount of information from the observable material. If the ancient record is to be obtained, the history must be stored in the minerals. Geophysical methods, which provide so much data about the present, can tell us less about the past. In this chapter we shall examine some mineral reactions and see how information is stored.

Today, we can reproduce earth conditions in the laboratory to a limited extent. Obviously, the immense accumulation of data from the 25°C–1 atm chemical laboratory must apply directly to the surface environment. But we are concerned with a body where P reaches millions of atmospheres and T about 4000°C. Our experimental knowledge of materials under these conditions is small and very inadequate to satisfy the potential applications.

Experimental apparatus is now available to carry out direct static experiments up to pressures of 10^5 bars and about 2000°C (note 1 bar = 0·987 atmospheres). Such conditions take us to depths of about 300 km. Gas-phase experiments rarely exceed 10^4 bars. In most cases P–V–T relations or phase changes are studied but a few measurements have been made of quantities such as electrical conductivity and optical absorption spectra.

Transient shock-wave experiments have been used to attain core pressures and temperatures. This type of experiment is of most value for P–V–T measurements in simple substances like pure metals. When complex polyatomic solids are involved, it is more unlikely that equilibrium states are attained during the microsecond times involved in shock experiments. There is a constant search for materials of higher strength for stronger test tubes.

For crustal chemistry, there is still a great lack of general data on the nature of solution systems at temperatures above about 300°C. This type of chemistry is vital to our understanding of many geochemical distribution processes; in particular those leading to the concentration of useful minor elements.

A constant problem arises in studying geological phenomena. Many important metamorphic and sedimentary processes are slow and occur over time periods of years or much longer. There is a tendency for these processes to reach equilibrium, but sometimes through a series of metastable intermediates. Laboratory times, particularly in very high P–T studies, are short. Frequently the states achieved experimentally are not the natural states, or, the natural states are only reproduced far from natural conditions. The experimentalist must be constantly aware of these limitations; he cannot afford to ignore the laws of chemistry or the observations of geology.

In principal, if we had adequate thermodynamic data (heats of formation, entropies, etc.) for all minerals, the battle would be largely won. But there are many minerals, the precision required for such data is very high, and few laboratories can produce calorimetric data with useful accuracy for our purposes. We must do experiments.

Solid–solid reactions: phase changes

Most minerals when heated or compressed undergo phase changes. The equilibrium phase diagram for a mineral or rock system is invaluable in discussing the history of the system. In this small book we can only outline a few typical cases.

Polymorphic changes are either of first order (there is a latent heat of transition and at the transition temperature C_p tends to infinity) or of second order (the C_p curve is lambda-shaped).

In Fig. 6.1 we show the phase diagram for the SiO_2. It is interesting for these polymorphs to compare entropy–density relations.

	Entropy/(J K^{-1} mol^{-1})	Density/(g cm^{-3})
Cristobalite	43·43	2·334
Tridymite	43·93	2·265
α-Quartz	41·34	2·648
Coesite	38·91	2·911
Stishovite	27·74	4·287

In general, for solid–solid transitions there is correlation between entropy and volume which is quantitatively rather constant. This means that the slopes of such first-order transitions on a phase diagram, given by the Clausius–Clapeyron equation†

$$\frac{dP}{dT} = \frac{\Delta S}{\Delta V}$$

tend to be similar. For most minerals made from oxide components, it is possible to estimate the entropy by summation of the entropies of the oxides and by allowing for a volume correction if necessary.

Calcium carbonate ($CaCO_3$) occurs in two modifications. The common mineral calcite occurs in a rhombohedral crystal form while a denser polymorph, aragonite, crystallizes in the orthorhombic system. Calcite has the larger entropy and volume and we would expect the phase relations as shown in Fig. 6.2. The difference in free energy of the two polymorphs at 25°C and 1 atmosphere is quite trivial.

$$G^{\ominus}_{\text{Aragonite}} - G^{\ominus}_{\text{calcite}} = \Delta G^{\ominus}_{\text{reaction}}$$

$$-(1128{\cdot}63)-(-1129{\cdot}60) = +0{\cdot}97\ \text{kJ mol}^{-1}.$$

The problem of obtaining significant thermodynamic data is obvious.

† The thermodynamic relationships in this chapter are discussed in E. B. Smith: *Basic chemical thermodynamics* (OCS 8)

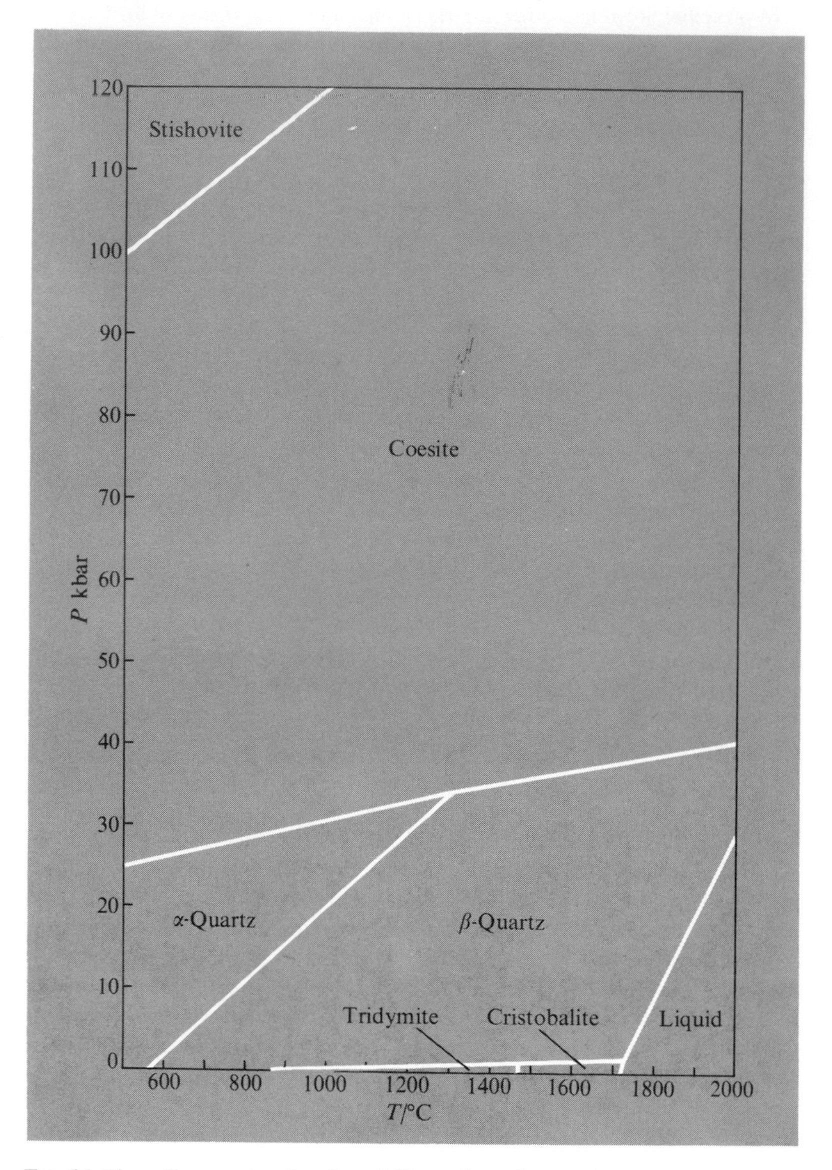

FIG. 6.1. Phase diagram showing the stability regions of some of the phases of formula SiO_2.

When calcium carbonate is synthesized in solution at low temperatures and one atmosphere pressure, either form can be obtained depending on solution chemistry, which influences nucleation kinetics. In fact, marine organisms grow either or both, presumably a function of their cell chemistry. But when temperatures over 100°C are attained, in geologic time, equilibrium tends to be obtained.

Thus when calcareous sediments containing both forms are buried, metastable aragonite is soon converted to calcite. In certain tectonic regions of the crust, calcite marbles are compressed to become aragonite marbles. They must follow a path such as A–B–C (Fig. 6.2) during burial. Now it will be noted on this diagram, that if the crustal thermal gradient was normal (say 20–30°C km^{-1}, P increases about 280 bars km^{-1}) then this transformation should never occur at all. The fact that it does tell us that, in this geological region, the thermal gradient is abnormally low. But we can go further. The kinetics of this reaction have been rather well studied and it has been shown that aragonite, formed at depth, could only reach the surface again it it returned along a path such as C–D–A (Fig. 6.2) where the thermal gradient is less than 10°C km^{-1}. It turns out that such a thermal gradient in the crust is possible only where material is rapidly descending into the mantle; a Benioff zone (p. 27). Thus a simple phase transition records a major tectonic event. In fact, we can go back

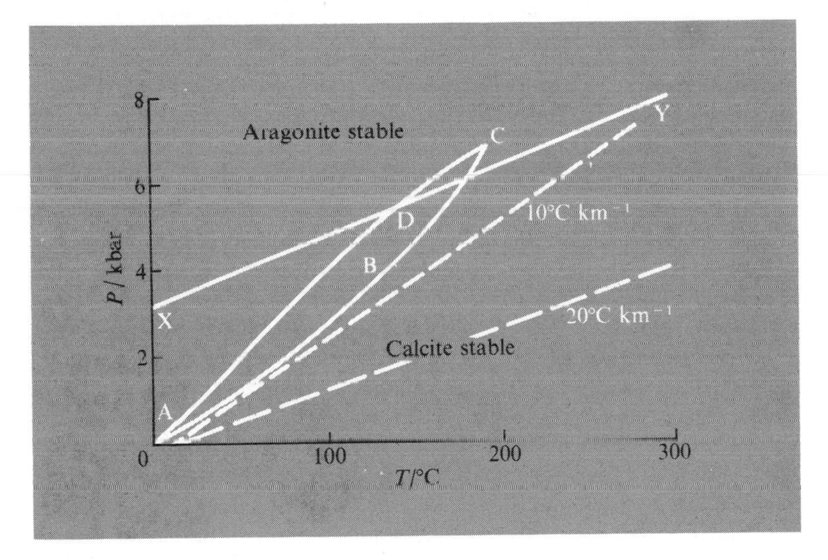

FIG. 6.2. Phase diagram for $CaCO_3$ showing the stability fields of the polymorphs calcite and aragonite. The dashed lines show P–T variation in the crust for thermal gradients of 10 and 20°C/km. For other details see text.

in time billions of years and use this type of information to record fossil convective structures.

The system Al_2SiO_5 has attracted much attention in relation to earth history and it is a system which clearly demonstrates the problems of obtaining a valid equilibrium phase diagram. Three common polymorphs exist and their basic thermodynamic properties are listed below:

	$V^{\ominus}/(\text{cm}^3\,\text{mol}^{-1})$	$S^{\ominus}/(\text{JK}^{-1}\,\text{mol}^{-1})$	$G^{\ominus}/(\text{J mol}^{-1})$
Kyanite	44·09	83·76	−2443 460 ± 2300
Andalusite	51·53	93·22	−2444 020 ± 3010
Sillimanite	49·90	96·11	−2441 780 ± 3010

The three polymorphs are wide spread in crustal metamorphic rocks being derived from the metamorphism of clay rich sediments. They have interesting structures, kyanite having all Al in six-coordination, sillimanite Al in six- and four- and andalusite Al in six- and five-coordination. The substances are highly refractory and in the dry state quite unreactive until temperatures above 1000°C even though we know from geological evidence that in rocks they form between 400–800°C. At high temperatures all polymorphs react to form the common refractory phase, mullite, and for the kyanite $\longrightarrow$ mullite transition, the activation energy has been found to be about 630 kJ mol^{-1}. This enormous activation energy attests to the stability of these structures.

From inspection of the data above, and the Clausius–Clapeyron equation, it follows that if these phases each have a region of stability, then the phase diagram must be as in Fig. 6.3. There have been dozens of attempts to determine these phase relations experimentally, and while today we have a rather good idea of where boundaries lie, the same diagram could have been produced from logical consideration of geologic maps in regions where transitions occur. What are the problems?

The free-energy data indicated above, based on calorimetric studies, are much too imprecise and because the substances are so reluctant to react is difficult to substantially improve. Only direct experiment will settle the problem. Near the transition temperature andalusite and sillimanite have almost identical entropies. This implies they are of almost identical stability and which phase nucleates and grows is a matter of many minor causes (on the laboratory timescale). The free energies are so similar that grain-size variation (and hence surface-energy effects) can change relative stabilities. The free energies are also so similar that strain energy (dislocations) induced by grinding can change relative stabilities, and tiny quantities of impurities can change stability. This is an extreme case but it illustrates the problem. Probably the best method to study relative stability in this system to date has involved the measurement of solubility of large strain-free crystals (in either fused-salt solvents or high P–T water). Geochemists have become very skilful at studying these complex systems and in producing very precise thermodynamic data. It should be

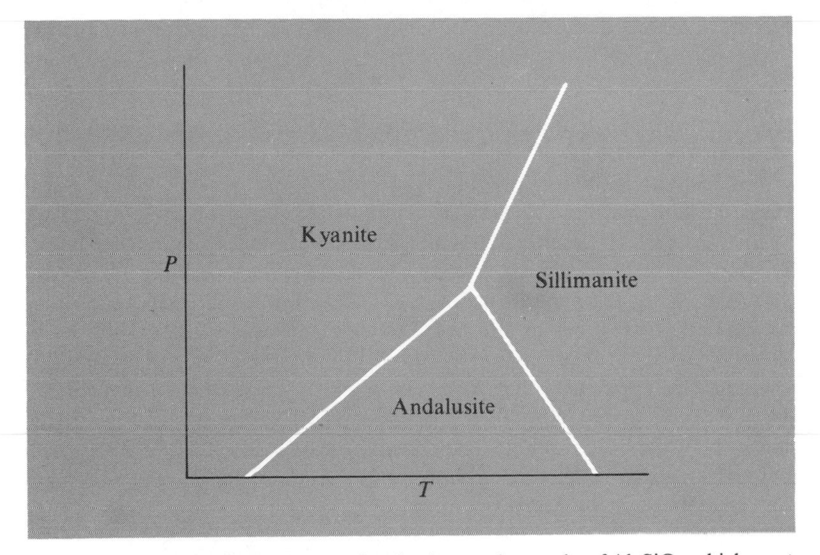

FIG. 6.3. Form of the phase diagram for the three polymorphs of Al_2SiO_5 which must follow from the thermodynamic variables.

remembered that a phase diagram is a statement of thermodynamic properties.

Polymorphs abound in nature and many have been studied in the laboratory. Gradually we can build up a grid of phase changes for the crust and mantle. In the mantle, regions of major phase changes (e.g. olivine → spinel, Mg_2SiO_4 → $(Mg_2Si)O_4$; or quartz → stishovite) correlate well with regions where seismic wave velocities show abnormal variation. In the crust, occurrence of given polymorphs indicates the possible *P–T* path to which a rock has been subjected. It is fortunate that in many cases, forms stable at high temperatures do not revert to the low-temperature form on cooling of the rock during return to the surface. We shall consider the reasons for this on p. 71. It should also be noted that at pressures near the core, even the electron distribution in atoms may change.

Following from the general principle that high-entropy states are favoured by high temperatures, a second major class of transitions occur which involve disordering. They are sometimes called lambda or second-order phase changes. They differ from the types we have discussed above in that no nucleation of a new phase is involved and there is no sharp transition temperature. The crystal quietly vibrates itself from one state through a continuous series of new states. There are many different examples of this type of phenomena in solids but all can be treated by the same thermodynamic approach involving statistical calculation of entropy of disordering from the Boltzmann equation

$$S = k \ln \Omega$$

where k is the Boltzmann constant and Ω the number of accessible quantum states for the system.

The feldspar minerals show this type of polymorphism. At low temperatures, the minerals albite $NaAlSi_3O_8$ or orthoclase $KAlSi_3O_8$ have framework structures in which the sites for Al and Si atoms are unique even though they are very similar and appear to play identical roles in building the crystal structure. As the temperature is raised and vibration amplitudes of the atoms increase, the two types of site begin to mix and above 1000°C the Al–Si distribution approaches a random state. The symmetry of the crystal also changes. The entropy of such a process is easily calculated. In the ordered form, we have three white (Si) and one black (Al) ball distributed in specific black and white boxes; at high temperatures the balls can still be distinguished but not the boxes. In the ordered state, $\Omega = 1$. In the random state, for a gram formula unit (L is Avogadro's number).

$$\Omega = \frac{(4L!)}{(3L!)(L!)}$$

and the ΔS of disordering is $2{\cdot}24R$ where R is the gas constant (note $k = R/L$).

For such cases the entropy of disordering is always positive but as the ordered structure is stable at low temperatures, the heat of disordering must be positive. The equilibrium state will be determined by $\Delta G = \Delta H - T\,\Delta S$ and eventually at some elevated T, the $T\Delta S$ term must dominate. But at any T above absolute zero, there must be a finite degree of disorder.

The feldspar case has been well studied and observations on site populations are in reasonable agreement with simple theory. Imagine the complexities in structures like the amphiboles (p. 19) where up to seven sites may be involved in a single crystal!

Other examples of this type of process important geologically include:

(1) Magnetic ordering in minerals like Fe_3O_4 which determines the temperature at which the mineral records the earth's magnetic field.

(2) Creation of defects in crystals (holes, interstitial ions, etc.) which determine the electrical conductivity of ionic crystals.

(3) Vibration-free rotation transitions in crystals involving non-spherical ions like the planar nitrate or carbonate ions.

More complex solid–solid reactions are common in all crustal rocks. We will consider an additional example here because it again illustrates how information comes from the study of rocks.

The very common feldspar mineral albite if highly compressed undergoes a reaction to form the pyroxene jadeite.

$$\underset{\text{albite}}{NaAlSi_3O_8} \rightarrow \underset{\text{jadeite}}{NaAlSi_2O_6} + \underset{\text{quartz}}{SiO_2}$$

The $\Delta V^{\ominus} = -17\ cm^3$ and $\Delta S^{\ominus} = -35{\cdot}1\ JK^{-1}\ mol^{-1}$ and the reaction essentially involves an Al in four coordination passing to a new structure with Al in six-coordination. The phase boundary for this reaction is shown in Fig. 6.4a.

In some rocks (metamorphosed andesites) nearly pure jadeite results from this transition but in general other pyroxene components enter into solid solution in the products. Thus the natural mineral is frequently a solid solution of $NaAlSi_2O_6$–$NaFeSi_2O_6$–$CaMgSi_2O_6$. Both $NaFeSi_2O_6$ and $CaMgSi_2O_6$ are quite stable at low pressures and by mixing with them, the high-pressure mineral $NaAlSi_2O_6$ is stabilized.

In natural rocks we often find a pyroxene solid solution in equilibrium with albite and quartz. We could study this particular rock experimentally and determine the conditions under which the observed solid solution is stable but as the range of solutions is so large we would like some method of extending more limited data. Obviously we must use solution theory.

Imagine we have an $NaAlSi_2O_6$–$NaFeSi_2O_6$ solid solution in equilibrium with albite and quartz at temperature T and we will assume that at T the

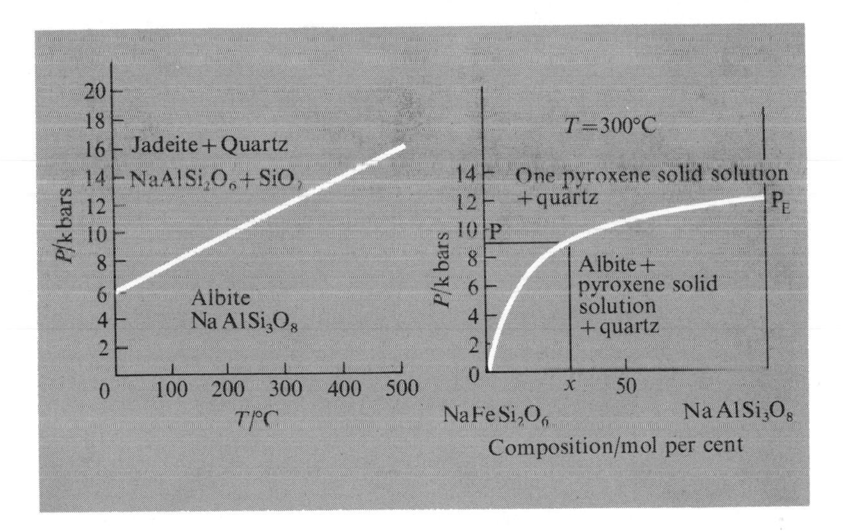

Fig. 6.4.(a) Phase diagram for the system $NaAlSi_3O_8$ showing the conversion of a feldspar to a pyroxene. (b) Phase diagram for an ideal solution of $NaAlSi_2O_6$–$NaFeSi_2O_6$ in the presence of quartz. For explanation see text.

pressure of equilibrium for the reaction albite → jadeite + quartz is P_E. The observed solid solution must form at a lower pressure. If the solid solution is ideal we can write the familiar equation relating chemical potential and concentration $\mu_{Jd,ss} = \mu^{\circ}_{Jd} + RT \ln x_{Jd,ss}$ where x_{Jd} is the mole fraction of $NaAlSi_2O_6$ in the solution. We know the composition of the mineral and hence x_{Jd}. We also have the equilibrium condition at P where the mineral formed that

$$\mu_{Jd,ss} + \mu_{Qtz} = \mu_{Albite}$$

or

$$\mu_{Jd} + RT \ln x_{Jd,ss} + \mu_{Qtz} = \mu_{Albite}$$

We can then write for pressure P and temperature T

$$\mu_{Jd} + \mu_{Qtz} - \mu_{Ab} = -RT \ln x_{Jd,ss}$$

and if the volumes of the phases change little with pressure (a reasonable assumption for most minerals),

$$(V_{Jd} + V_{Qtz} - V_{Ab})(P_E - P) = -RT \ln x_{Jd,ss}$$

remembering that

$$\left(\frac{d\mu}{dP}\right)_T = V.$$

As x is known and P_E is known we have the *exact* P at which the mineral formed. For a constant temperature, the phase diagram for this system is as shown in Fig. 6.4b.

Experimental studies show that this system closely approaches ideality (partly resulting from the fact that Fe^{3+} ion, $3d^5$, shows no crystal-field stabilization effects). This type of equilibrium abounds in rocks and if we have another such equilibrium in the same rock and mineral assemblage, we could fix both P and T simultaneously. In fact, the time is coming when we can examine a rock and say that it formed at $P_x T_x$ (depth related to P_x) which is one of the most valuable pieces of information we can have. Such data record the geothermal gradient and hence the tectonic situation of the rock. The same data provide information on the physical state of fluids which may be active in the region because the solvent power and solute chemistry of high P–T fluids are much influenced by these variables (p. 79).

It is interesting to note that the reaction albite → jadeite + quartz has not been recorded in rocks older than 10^9 years. Either thermal gradients were too high or the type of convection occurring in past times did not drag appropriate lighter materials to sufficient depth.

Caution is necessary. Solutions are not always ideal and in fact regular solution models are superior for most examples of mineral equilibria. It is also necessary to have data on the types of lattice sites involved in mixing, because this will determine the form of the ideal solution equation. Obviously mixing $NaAlSi_2O_6$–$NaFeSi_2O_6$ involves one lattice site per mole. But mixing $NaAlSi_2O_6$ and $CaMgSi_2O_6$ is different and the entropy of mixing must be doubled. In dealing with minerals one must always keep in mind that the relation:

$$\mu_i = \mu_i^{\ominus} + RT \ln x_i \text{ can be derived from } S = k \ln \Omega.$$

If solid solutions abound in rocks, it is obvious that every time a new phase is generated, elements are redistributed among the appropriate sites of each mineral of the rock assemblage. Geochemists are much concerned with such distribution coefficients for they finally influence the total behaviour of a minor or major element during mineral reactions. Most minor elements are tucked away in some suitable lattice site in a common mineral of a given rock. But at times there is no suitable phase, and either a separate mineral of the rare element will occur or the element may be swept out of the rock in a fluid phase. The subject is complex and again a realistic understanding hinges on mixing and solution theory.

Solid–gas reactions

Sediments and other rocks formed near the surface of the earth tend to be hydrated and oxidized (p. 28). If these sediments are buried and metamorphosed, perhaps the dominant process occurring is progressive loss of water, and with carbonate sediments, loss of carbon dioxide. The water in its passage to the surface transports soluble materials, often forming valuable new deposits (Chapter 7). The quantities of water involved are quite vast.

Typical of these dehydrations are:

$$\underset{\text{kaolinite}}{Al_4Si_4O_{10}(OH)_8} + \underset{\text{quartz}}{SiO_2} \rightarrow \underset{\text{pyrophyllite}}{2Al_2Si_4O_{10}(OH)_2} + \underset{\text{water}}{2H_2O}$$

$$\underset{\text{pyrophyllite}}{Al_2Si_4O_{10}(OH)_2} \rightarrow \underset{\text{kyanite}}{Al_2SiO_5} + \underset{\text{quartz}}{3SiO_2} + \underset{\text{water}}{H_2O}$$

$$\underset{\text{analcime}}{NaAlSi_2O_6 \cdot H_2O} + \underset{\text{quartz}}{SiO_2} \rightarrow \underset{\text{albite}}{NaAlSi_3O_8} + \underset{\text{water}}{H_2O}$$

$$\underset{\text{muscovite (mica)}}{KAl_2(AlSi_3O_{10})(OH)_2} \rightarrow \underset{\text{orthoclase}}{KAlSi_3O_8} + \underset{\text{corundum}}{Al_2O_3} + \underset{\text{water}}{H_2O}$$

$$\underset{\text{calcite}}{CaCO_3} + \underset{\text{quartz}}{SiO_2} \rightarrow \underset{\text{wollastonite}}{CaSiO_3} + \underset{\text{carbon dioxide}}{CO_2}$$

$$\underset{\text{paragonite}}{NaAl_2(AlSi_3O_{10})(OH)_2} + \underset{\text{sodium chloride}}{2NaCl} + \underset{\text{quartz}}{6SiO_2} \rightarrow \underset{\text{albite}}{3NaAlSi_3O_8} + H_2O_{gas} + 2HCl_{gas}$$

$$\underset{\text{hematite}}{Fe_2O_3} + \underset{\text{quartz}}{SiO_2} \rightarrow \underset{\text{fayalite}}{Fe_2SiO_4} + \tfrac{1}{2}O_{2,\,gas}$$

$$\underset{\text{pyrite}}{FeS_2} \rightarrow \underset{\text{pyrrhotite}}{FeS} + \tfrac{1}{2}S_{2,gas}$$

Understanding these gas-forming reactions is absolutely critical to our knowledge of fluid chemistry in the crust.

The equilibrium relations of many are now well known from experimental studies and all can be approached by the same thermodynamic methods. At low pressures and temperatures, most gases are sufficiently ideal that $PV = RT$ can be applied to the gas phase. At elevated temperatures (say $>600°C$) this equation is moderately satisfactory for aqueous fluids to pressures of 1000 atmospheres or so.

For any equilibrium of the type

$$AB_{solid} \longrightarrow A_{solid} + B_{gas}$$

we can write

$$G_{AB} = G_A + G_{gas}$$
$$G_{AB} = G_A + G^{\ominus}_{gas} + RT \ln P_{gas}.$$

At low pressures where gas volumes are much larger than solid volumes we can write the approximation:

$$\Delta G^{\ominus}_{reation} = -RT \ln P.$$

At high pressures we cannot use such a simple relation because the gases are not ideal and volume terms for the solids are significant. Thus if for temperature T we have thermodynamic functions for all substances at 1 atmosphere. In general at equilibrium:

$$G_{reactants} = G_{products}$$

$$G^{\ominus}_{reactants} + \int_{P=1}^{P_{equil}} V_{reactants}\, dP = G^{\ominus}_{products} + \int_{P=1}^{P_{equil}} V_{product}\, dP$$

and

$$\Delta G^{\ominus} = \int_{P=1}^{P_{equil}} \Delta V\, dP$$

and this relation must be integrated to find the equilibrium conditions.

In the crust with its complex chemistry, we have an almost continuous series of dehydration steps between the surface down to regions where melting occurs. For a simple dehydration series (like kaolinite above) where no solid solutions are involved, it is normal to find that in reactions like:

$$A{\cdot}3H_2O \longrightarrow A{\cdot}2H_2O + H_2O \quad (1)$$

$$A{\cdot}2H_2O \longrightarrow A{\cdot}H_2O + H_2O \quad (2)$$

$$A{\cdot}H_2O \longrightarrow A + H_2O \quad (3)$$

that $\Delta H_3 > \Delta H_2 > \Delta H_1$ while $\Delta S_3 \sim \Delta S_2 \sim \Delta S_1$. These relations mean that dehydration curves or vapour-pressure curves will have a form as in Fig. 6.5. Two typical vapour-pressure curves for mica-type phases are shown in Fig. 6.6.

As rocks are progressively heated, common minerals loose water in a sequence: first, clays and zeolites (around 300°C); second, chlorites (around 400°C); third, micas and amphiboles (600–1000°C).

An interesting and significant vapour-pressure curve is shown by the reactions of the zeolite analcime:

$$\underset{\text{analcime}}{NaAlSi_2O_6{\cdot}H_2O} + \underset{\text{quartz}}{SiO_2} \rightarrow \underset{\text{albite}}{NaAlSi_3O_8} + H_2O$$

The vapour-pressure curve (Fig. 6.7) is a loop and it will be noticed that at constant T, when water pressure is increased, we can pass from anhydrous solids → hydrous solids → anhydrous solids; dehydration is caused by increasing water pressure. The necessary thermodynamic condition for this process is that

$$V_{\text{anhydrous product}} + V_{H_2O} < V_{\text{hydrate}}$$

and this is true for the above system at all moderate pressures.

While few examples of this type of loop have been quantitatively studied, it is certain that at upper mantle pressures the above condition is appropriate

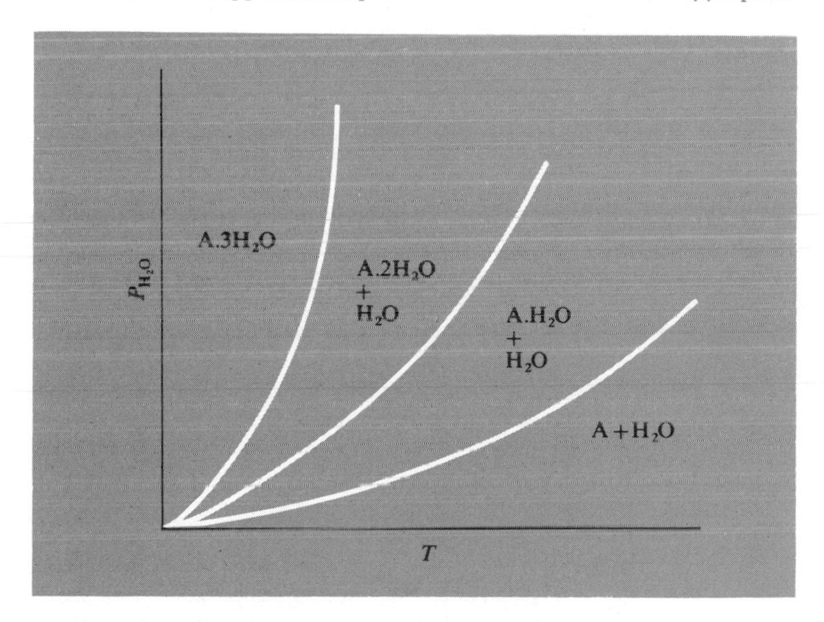

FIG. 6.5. 'Normal' form of a series of dehydration curves where $\Delta H_3 > \Delta H_2 > H_1$ etc. while $\Delta S_1 \simeq \Delta S_2 \simeq \Delta S_3$ etc. (see text).

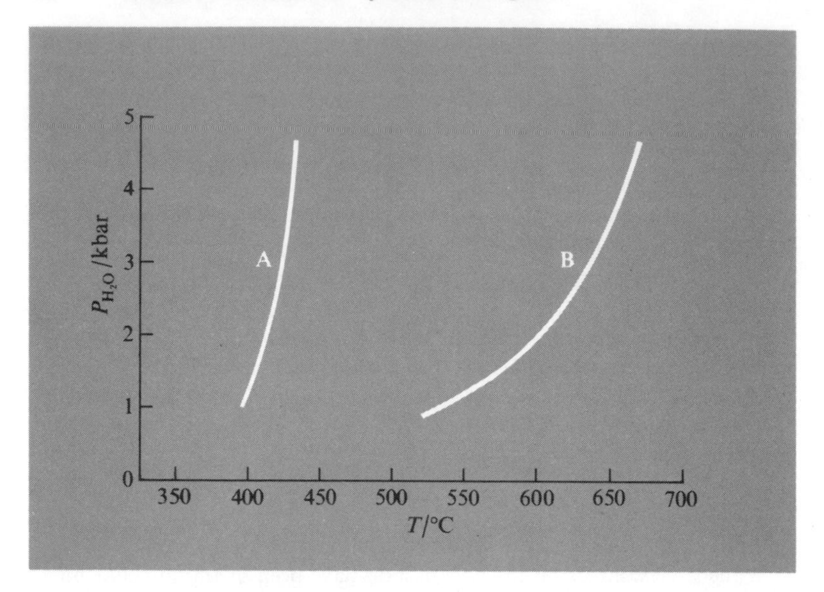

FIG. 6.6. Two typical vapour-pressure curves for common metamorphic reactions: A for the breakdown of pyrophyllite:

$$Al_2Si_4O_{10}(OH)_2 \rightarrow Al_2SiO_5 + H_2O;$$

B for the breakdown of muscovite

$$KAl_2AlSi_3O_{10}(OH)_2 \rightarrow KAlSi_3O_8 + Al_2O_3 + H_2O$$

to almost all known hydrated mineral phase. Is there water in the deep mantle? If so we know little of how it is chemically bound (we can make some guesses) and certainly the states are not familiar to us.

Reactions producing other gases have also been studied. With carbonates, some sulphides, and halide systems, quite large partial pressures of the reactant gases may result at moderate temperatures. With common oxidation–reduction systems the partial pressures of oxygen are much lower. In Fig. 6.8 we show the vapour-pressure curves for some reactions involving the system Fe–O. A pair of minerals such as Fe_3O_4–Fe_2O_3 effectively buffer the oxygen pressure of an environment. This buffer action is quite important in controlling solubilities of some materials like gold (p. 80). At depth, in rocks where there is no chance of atmospheric interaction, the most common values of P_{O_2} lie near the curves for the fayalite–magnetite reaction (Fig. 6.8).

The partial pressure of oxygen, arising from the self-dissociation of H_2O, $2H_2O \rightarrow 2H_2 + O_2$, is larger than for these iron systems under most crustal conditions. This means that natural aqueous systems at depth will contain an excess of hydrogen over oxygen because reactions like

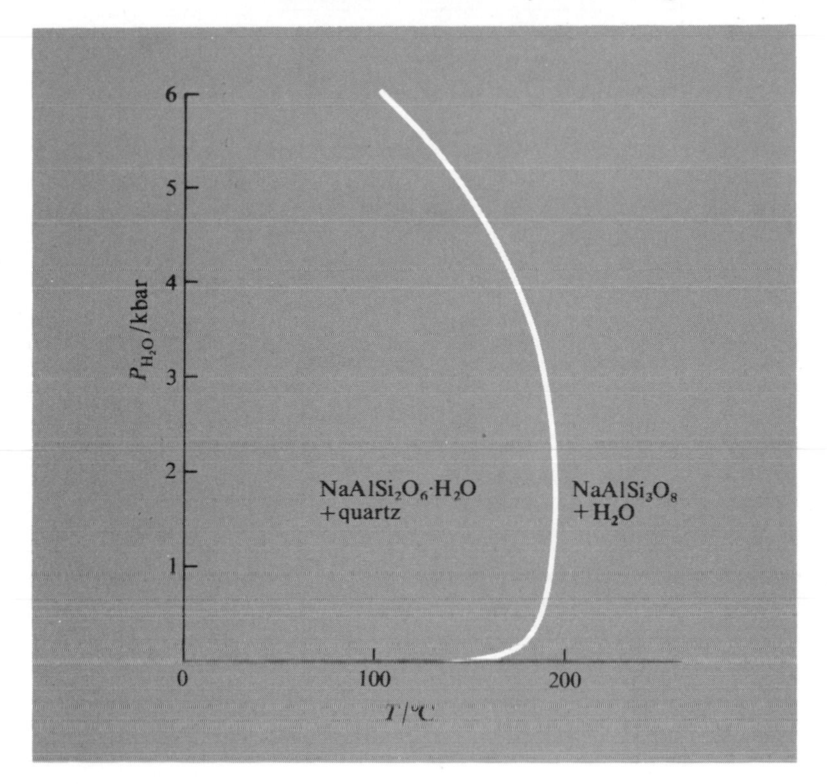

FIG. 6.7. Vapour pressure curve for the reaction: $NaAlSi_2O_6{\cdot}H_2O+SiO_2 \rightarrow NaAlSi_3O_8+H_2O$. Note, at low pressures, the curve reaches the origin at 0 K. It cannot be shown on this scale.

$$2Fe_3O_4 + H_2O \rightarrow 3Fe_2O_3 + H_2$$

tend to occur until an equilibrium P_{H_2} is reached. This is an important result because it means that fluids from depth coming to the surface add hydrogen, not oxygen, to the atmosphere. If hydrogen did not escape into space we should have a reducing atmosphere (ignoring biological processes).

Isotope fractionation

The different nuclear properties of isotopes lead to slight variation in the thermodynamic functions for each isotopic species. These differences lead to slightly different equilibrium conditions (e.g. boiling points of compounds, etc.) for each isotopic species and hence in a general way, to isotope fractionation during any chemical process.

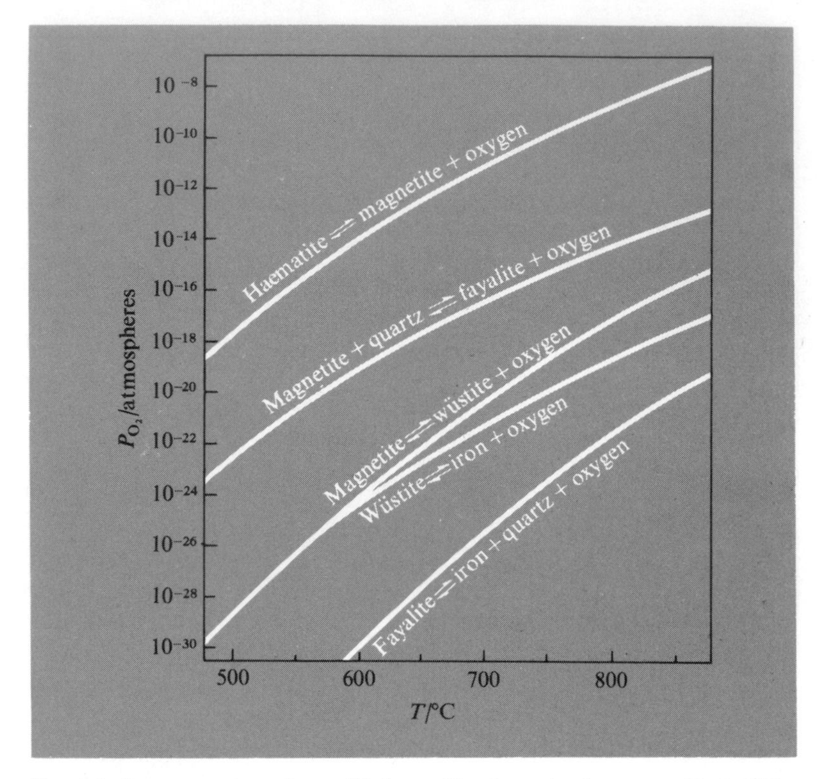

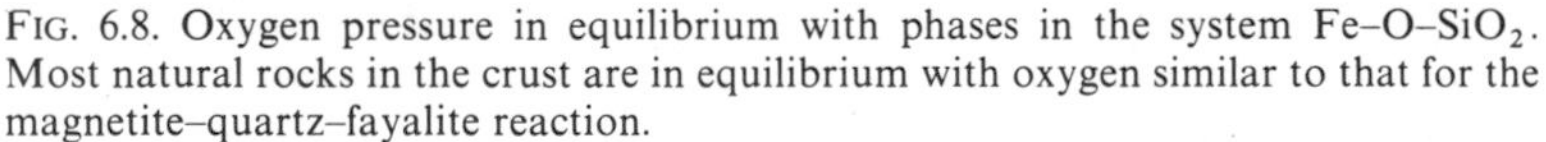
FIG. 6.8. Oxygen pressure in equilibrium with phases in the system Fe–O–SiO_2. Most natural rocks in the crust are in equilibrium with oxygen similar to that for the magnetite–quartz–fayalite reaction.

In natural systems such fractionation is most apparent when light isotopes are involved in low temperature processes. In nature, significant fractionation has been observed with hydrogen–deuterium, carbon, oxygen, and sulphur isotopes. Significant fractionation with elements beyond sulphur has not been observed. We shall here use oxygen isotope fraction $^{16}O:^{18}O$ to indicate some of the information that can be obtained from fractionation measurements.

Natural surface waters differ considerably in their oxygen isotopes. Light water $H_2\,^{16}O$, has a higher vapour pressure than $H_2\,^{18}O$, and so is concentrated by evaporation processes. Thus fresh water is generally lighter than ocean water and glacier or polar ice lighter still. When minerals like $CaCO_3$ or SiO_2 precipitate from such waters they are enriched in the heavy isotope and thus marine carbonates are heavier than fresh-water carbonates.

The extent of fractionation may be part kinetically controlled but often shows an approach to equilibrium. If equilibrium is attained, fractionation is

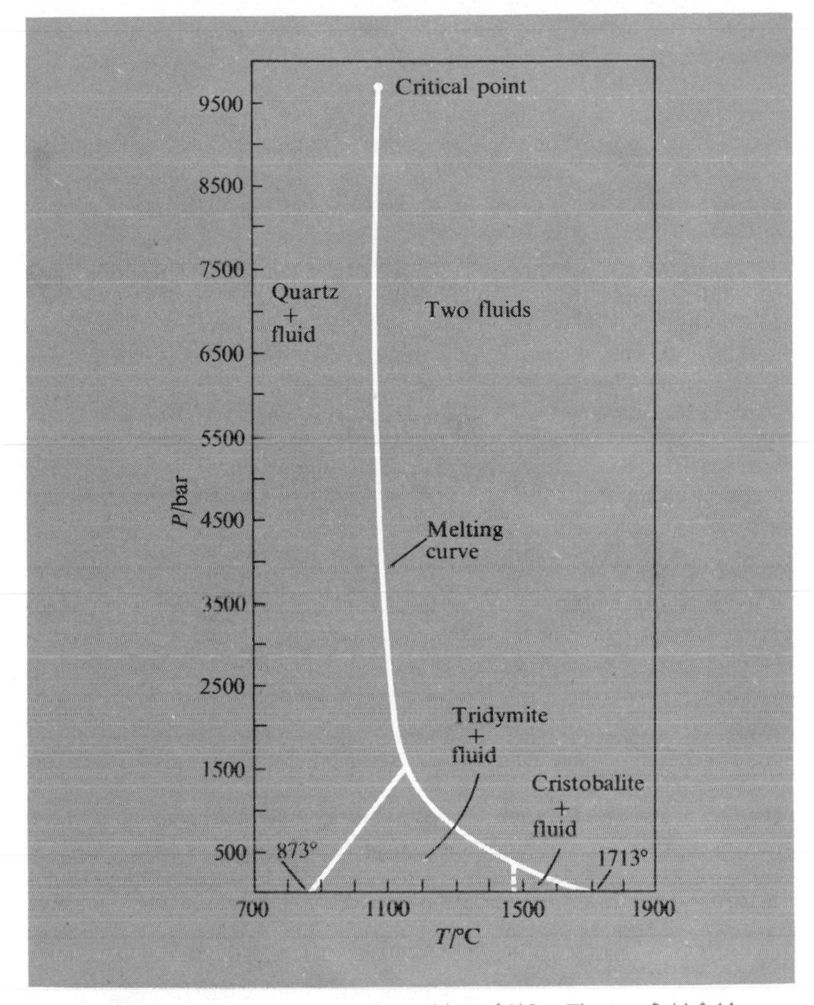

FIG. 6.9. Effect of water pressure on the melting of SiO_2. The two-fluid field represents coexistence of a gas rich in SiO_2 and a melt rich in water.

a function of temperature and if we grow $CaCO_3$ from water it will become lighter as T increases. Measurement of such ratios in carbonates precipitated by marine organisms has been used to estimate past ocean temperatures, often with a remarkable degree of success.

Thus, study of the isotopes can tell us something about the type of water in an environment and the temperature of interaction. While many rocks in the crust crystallize and recrystallize in the presence of water, we cannot observe the transient aqueous phase. But if two minerals containing oxygen crystallize in equilibrium with the common oxygen reservoir of the rock system, they will fractionate the isotopes and this fractionation will depend on T only. Most crustal rocks contain mineral pairs like quartz–magnetite (Fe_3O_4), quartz–albite, which show sufficient fractionation to make them useful thermometers. This is now one of the most powerful and simple tools we have for determining the temperature at which a rock formed. Naturally, the calibration of the fractionation constants must be carried out in the laboratory.

This method of finding temperatures naturally becomes less sensitive as temperature increases. In addition, for a record to be left in two associated minerals formed at an elevated temperature, it is necessary that interdiffusion of oxygen between the solids does not occur during the later history. But diffusion coefficients in solids are small (p. 71) and this method of thermometry may be generally useful to temperatures of 500–600°C and in special cases, when a short-term phenomenon operates, to even higher temperatures. The potential uses of stable isotope fractionation are very great indeed and increasing study will add much to our geochemical knowledge over the next decade.

Solid–melt equilibria

Heat input into any given region of the earth may eventually raise the temperature to a point where melting commences. As rocks are mineralogically and chemically complex, melting is normally a gradual process and earth materials have a melting range. When melting commences, further heat input contributes to the latent heat of melting of the region (normally of the order 400 kJ kg^{-1}) and once sufficient low-density melt is formed it may rise, carrying heat and mass to higher levels in the earth. This type of process occurring at any depth, forms the main thermal buffer of the earth.

There are three simple types of melting situation relevant to earth materials. With most normal solid materials melting temperatures increase with pressure. This is why the earth's inner core is solid yet must be hotter than the liquid outer core. Melting is normally a positive-entropy volume process and hence dP/dT must also be positive. If we melt a system consisting of solid + gas, if the gas is soluble in the melt, the melting point of the solid will be depressed by the formation of a solution; this is the normal freezing-point depression situation. The melting temperatures of most silicates are strongly depressed by excess water at high pressures. Thus the melting point of pure SiO_2 (1713°C) is depressed to 1100°C by about 1500 bars of water pressure (Fig. 6.9). It should be noted that these curves, at least at low pressures, have negative slopes; the entropy change is still positive, but the ΔV of solution of low-pressure gas in the melt is negative and hence dP/dT must also be negative. At very high pressures, where the densities of compressed gases approach those of solids, melting again tends to increase with pressure. Melting behaviour in situations where unlimited water is available to saturate a melt are uncommon in natural systems. But there are some small, economically significant examples, where water-saturated melts may be formed.

The final common type of melting behaviour occurs where the vapour-pressure curve of a hydrated mineral of a rock assemblage, intersects the melting curve of the system. The situation is shown in Fig. 6.10. In this case, the melting reaction is:

hydrate + A + B + C... $\rightarrow$ partial melt + X + Y + Z ....

Again this is a positive ΔS and ΔV process and the melting temperature increases with pressure. Most melting in the crust of the earth is probably of this type, for we know that the regions of stability of common high-temperature hydrated metamorphic minerals such as biotite and hornblende intersect the region of melting of average crustal materials.

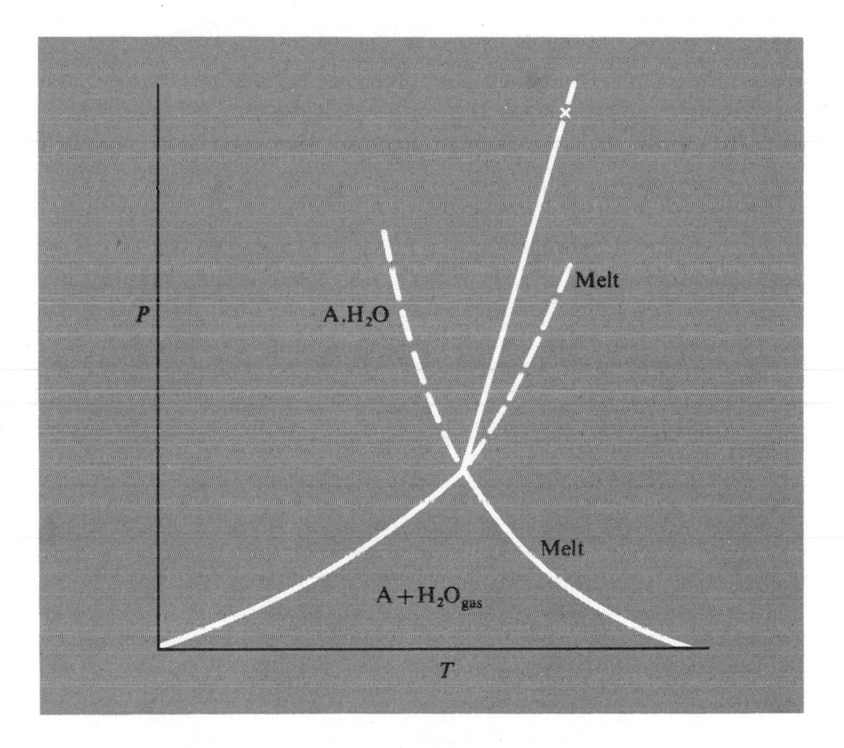

Fig. 6.10 Schematic representation of the melting of a hydrate where the vapour pressure curve of the hydrate intersects the depressed melting curve. This type of behaviour is probably very common in the base of the crust and upper mantle.

In modern volcanic regions as indicated on p. 27 we have noted three main types of melts; basaltic, andesitic, granitic. Their melting temperatures are in the order

$$\text{basaltic} > \text{andesite} > \text{granitic.}$$

Melting temperatures tend to decrease with increasing amounts of SiO_2, $Na_2O + K_2O$, etc. and increase with increasing amounts of MgO, FeO, or CaO. In ancient rock series, there is evidence that much higher temperature melts appeared from the mantle and produced rocks essentially made of two minerals, olivine and pyroxene. The rock is termed a peridotite. It should be emphasized that if melting effects a limited-volume region of the earth, the fractional melts formed will be in a strict thermal order. At a given depth we could not produce basalts first and granites second by fusion; only the reverse. It is obvious that volcanic rock series must provide evidence about what is melting and where melting is occurring.

If a melt is formed on a positive melting curve (say point X in Fig. 6.10) and rapidly rises to the surface, it moves into the field of liquid as it travels. Some melts arrive at the surface and are quenched to almost perfect glass. If the walls along the path of travel are incorporated, small amounts can dissolve in the melt without cooling to the point of crystallization; the process is called assimilation.

If a partially-wet melt rises, at low pressures, gases will nucleate and bubbles form in the melt. Some of the most spectacular and disastrous volcanic eruptions involve explosive nucleation caused by fast pressure release. Hundreds of km^3 of silicate froth may be erupted in a short period of time, blanketing the surroundings in glassy volcanic ash. This gas–glass mixture can flow for tens of miles from a source and can also contribute large quantities of dust to the upper atmosphere.

Melt–solid interactions

Complex fractional melting or fractional crystallization processes which occur with complex rocks can be correlated with behaviour in certain simple systems. If two substances melt to form totally miscible liquids, and if the solids are immiscible, a simple-eutectic melting pattern results (Fig. 6.11). The melting diagrams can be described by the simple thermodynamic conditions that at any temperature:

$$\begin{aligned} \mu_{A,\,solid} &= \mu_{A,\,liquid} \\ &= \mu^{\ominus}_{A,\,liquid} + RT \ln x_{A,\,liquid} \end{aligned}$$

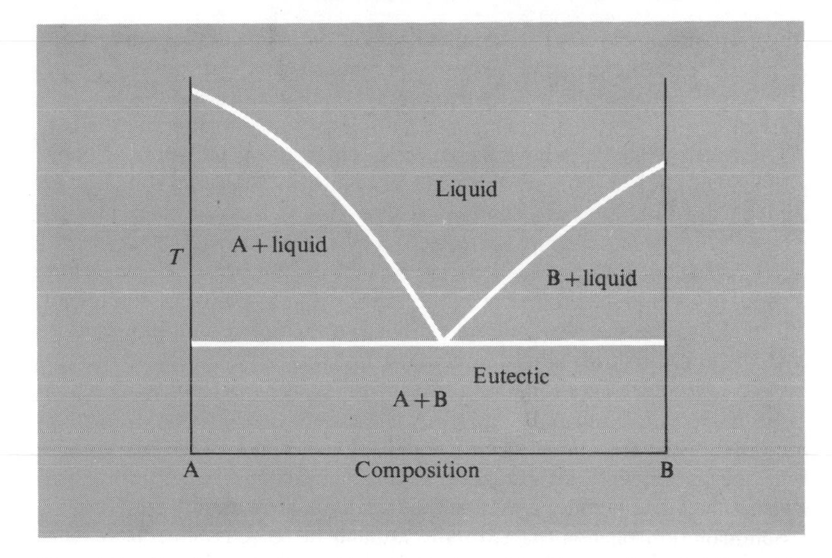

FIG. 6.11. Simple eutectic melting behaviour. Liquids totally miscible, solids immiscible.

This equation for ideal mixing in the liquid state leads to the equation for freezing point depression:

$$\ln\frac{1}{x_A} = \frac{\Delta S_A}{R}\left(\frac{T_m}{T} - 1\right),$$

where ΔS_A is the entropy of fusion of pure A and T_m is the melting temperature of pure A.

In such a simple system it is possible to crystallize pure A or B only over a limited temperature range and final crystallization of the eutectic liquid will produce a mixture of the two phases. In a gravitational field, if the cooling time and melt viscosity permits, a layer of a nearly pure phase can separate in such a system. The effect is commonly seen in igneous rock masses where separating phases may either sink or float in a magma chamber often leading to remarkably clean separations. Such separations can lead to valuable mineral deposits. In the spectacular Bushveld complex of South Africa for example, a mass of essentially basaltic magma crystallized slowly and produced a layered series of minerals. In a region near the base chromite ($FeCr_2O_4$) and platinum were deposited. The platinum-bearing layer is about 12 inches thick and has been followed for tens of miles. It is one of the major platinum occurrences of the world. The same separation processes form the chromite deposits which supply most of the world's chromium for stainless steel.

It is also obvious that, if a mixture of A + B is subjected to fractional fusion, the first melt will always have the eutectic composition. This eutectic composition will change with pressure and will change with the addition of other components, but the broad similarity of the composition of many magma types suggests that the simple eutectic model is often near the truth.

Solids do not always melt congruently and incongruent melting occurs with many compounds. This behaviour is shown in Fig. 6.12. A compound A_3B melts to form A plus a liquid enriched in B. If for example, a melt of composition X is slowly cooled, A is the first phase to crystallize and does so while the liquid changes composition along the path X_1–X_2. At X_2, A reacts with the liquid to form A_3B until all A is removed and then crystallization proceeds to the eutectic E where solidification occurs. Again this type of behaviour is frequently seen in igneous rocks; a primary phase is resorbed at a later stage or sometimes it may sink out of the system and be preserved.

Solid-solution phenomena naturally abound in igneous mineral systems where T is high and the effects of the positive entropy of mixing are maximal.

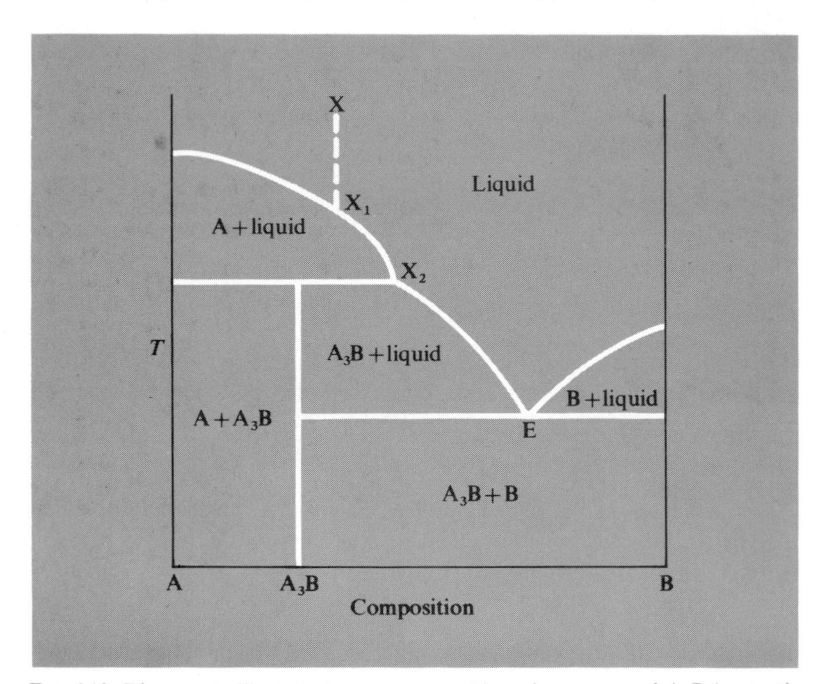

FIG. 6.12. Diagram to illustrate incongruent melting of a compound A_3B (see text).

Behaviour of the type shown in Fig. 6.13 is very common where both liquid and solid phases are totally miscible. If a liquid of composition X crystallizes the first crystalline phase will have composition X_1. If equilibrium obtains, as the system cools, the liquid will change composition to X_2 while the solid continuously reacts with the liquid to produce a solid of composition X when crystallization will cease. If cooling is fast or if early crystals sink out of the region of interaction, zoned crystals will result where a central core of high melting material is covered with successive layers of lower-melting solid. The liquid composition can now go beyond X_2 and nearer the origin.

Add these three types of behaviour together, add departures from perfect equilibrium, add gravitational settling, and add the possibility of solution of the container walls, and it is not difficult to see that the range of liquid compositions seen at the surface can be almost infinite. In general almost every batch of a magma in a volcanic region will have unique chemistry. Archaeologists have long used this to trace the source of artifacts made from volcanic materials.

Finally we may take two rather simple systems to indicate the complexity of happenings during crystallization of a magma. But these complexities make good thermodynamic sense. First, consider the crystallization of a feldspar

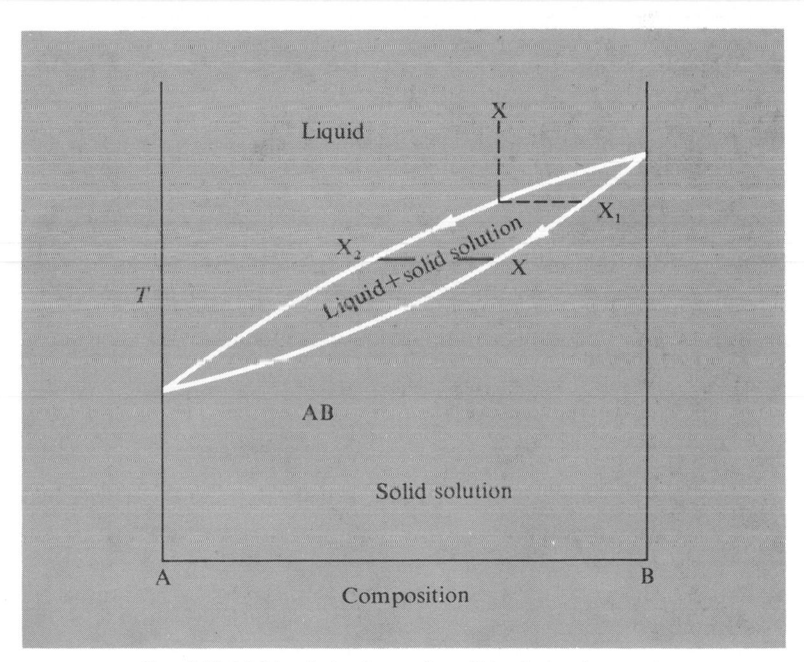

FIG. 6.13. Melting behaviour of a solid solution (see text).

in the system $KAlSi_3O_8$–$NaAlSi_3O_8$, a system common in all granitic rocks (Fig. 6.14). It will be noted:

(a) $KAlSi_3O_8$ melts incongruently to produce a phase leucite, ($KAlSi_2O_6$) + liquid.
(b) $KAlSi_3O_8$–$NaAlSi_3O_8$ at moderate temperatures form a continuous range of solid solutions but with a minimum melting point.
(c) At low temperatures, the solid solution unmixes.

Let us take a liquid of composition X and follow its crystallization. X will cool to X_1 where leucite will commence to crystallize; when the temperature reaches T_1, feldspar appears and leucite begins to be resorbed and when temperature T_2 is reached leucite disappears and a liquid X_2 is in equilibrium with a solid solution of composition X_3. With further cooling, the liquid moves to X_5 where it is in equilibrium with solid solution X_4; at which point

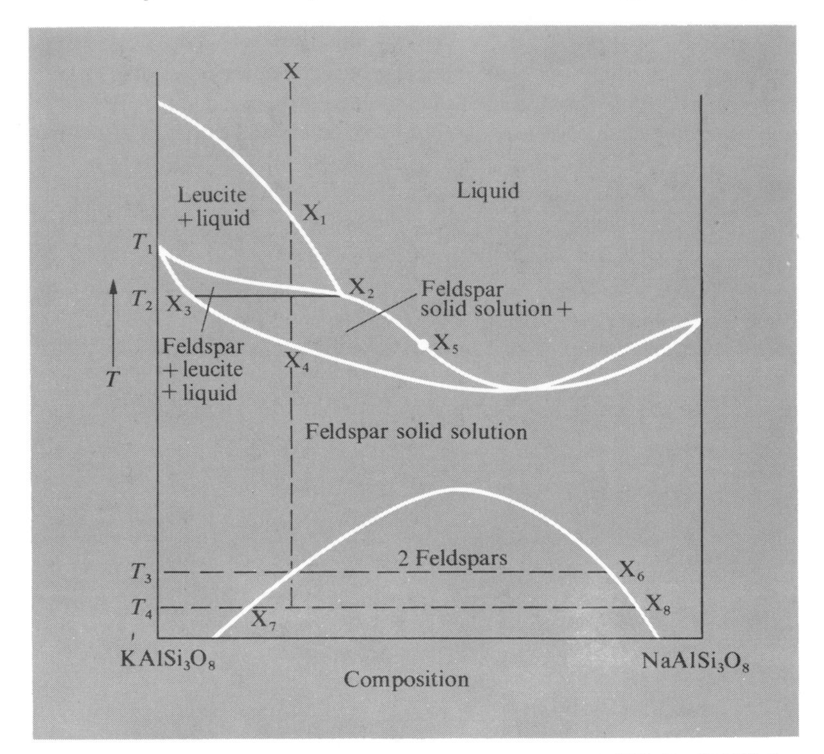

FIG. 6.14. Schematic diagram of melting in the feldspar system $NaAlSi_3O_8$–$KAlSi_3O_8$. This system exhibits incongruent melting, solid solutions with a minimum melting point, order–disorder transitions in the solids, and unmixing of a solid solution (see text).

the system becomes all solid. Solid solution X_4 now cools to T_3 and during this path ordering of Si–Al atoms in the feldspar sets in. At T_3, nucleation of a new phase (X_6) occurs by exsolution* and with further cooling the two solid phases increase in purity and degree of ordering. At T_4, our mixture consists of two feldspars X_7 and X_8. This type of exsolution occurs with most non-ideal solid solutions (ΔS_{mixing} + ve, ΔH_{mixing} + ve, ΔG_{mixing} + ve or − ve depending on T) and particularly where ions have significantly different radii. The rather sloppy high-T crystal lattice will tolerate the two types of ions (K^+, radius 1·33 Å; Na^+, radius 0·98 Å), but at low Ts where the amplitude of vibration of atoms is smaller, unmixing occurs to eliminate lattice strain, which would cause a reduction in lattice energy.

As igneous rocks are the primary materials of the earth's crust, geochemistry is much concerned with what happens to all elements during melting and crystallization. The major elements form the dominant minerals. The fate of minor elements is highly varied. If a minor element forms a very high melting compound, this implies it may be quite insoluble in a melt and crystallize as a separate phase at an early stage of magmatic crystallization. Thus the melting temperature of common basalt is about 1200°C. The basalt–chromite system may appear as in Fig. 6.15, the eutectic being very close to the first crystallization temperature; and even if the concentration of Cr_2O_3 in the melt is small, it may appear as an early phase during crystallization.

In general, the type of behaviour shown by Cr is not common. Transition metals like Ni, V, Mn, etc., are accommodated as solid solutions in the common olivines and pyroxenes. Elements like Rb are camouflaged in potassium minerals. Ionic radii and general atomic behaviour provide a good indication of where the rare element may be hiding. But if the element has an unusual radius (very large Cs^+, or small Be^{2+}, Li^+), or has odd oxidation states (W–U–Sn), *and* forms low melting-temperature compounds, then such an element may accumulate in the final fractions to crystallize. The same final-melt fraction may be rich in volatile non-metals (CO_2, H_2O, HF, HCl, B_2O_3, etc.) and these liquids often provide major sources of compounds of some rare metals. The melting temperatures of these late, volatile, saturated melts may be of the order of 600°C. They form the so-called pegmatites, characterized by large crystals (often metres long) and from them come our sources of many odd metals such as Li, Cs, U, Sn, Mo, W, Ta, etc.

Geochemists have long searched for rules, generally based on ionic radii or electronegativities, to explain the order of crystallization of the elements in igneous rocks. The rules have in general been of dubious success, and one need not look far for the reasons. Crystallization or melting is a solid–liquid equilibrium. If attention is focussed essentially on lattice-energy considerations for the solid, obviously we must be ignoring half the story, for the 'lattice energy' of liquids is almost as large as solids (i.e. heats of fusion are normally only a per cent or less of lattice energies).

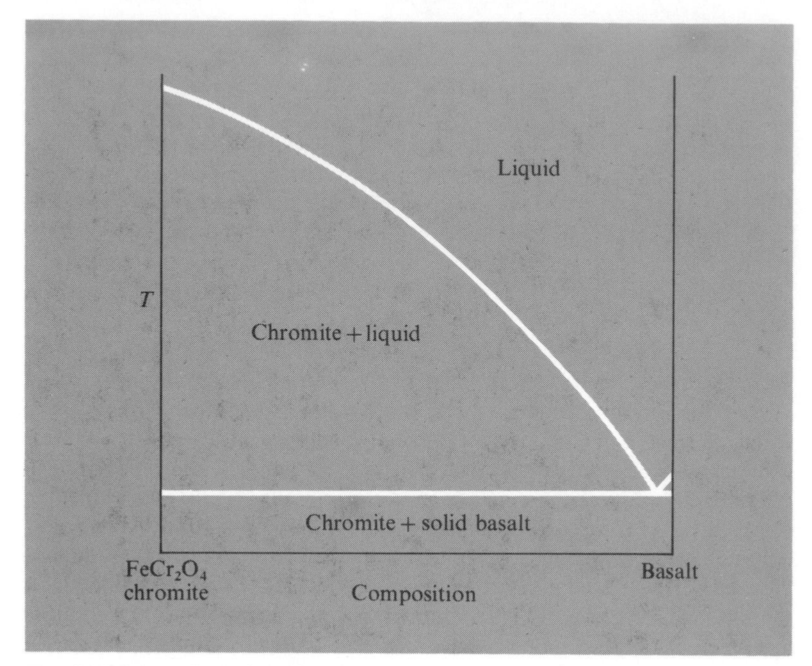

FIG. 6.15. A hypothetical melting diagram for the system chromite–basalt. Although the amount of Cr_2O_3 in a system may be very small (almost pure B) chromite may appear as an early solid phase to crystallize.

A nice example of these complexities is provided by the behaviour of nickel in melts. Basaltic rocks contain about 160 p.p.m. of nickel. When a basaltic melt crystallizes most of this nickel appears in solid solution in olivine (i.e., Ni_2SiO_4–Mg_2SiO_4). The first olivines to crystallize are richer in nickel than later olivines so that in basalts, the system Ni_2SiO_4–Mg_2SiO_4 appears to behave as in Fig. 6.16a). But when we study the simple olivine system we find exactly the inverse behaviour in that early olivine is enriched in magnesium (Fig. 6.16b). Clearly this inversion of behaviour is not a function of the solids but of changes in the liquid.

Studies of the structures of silicate liquids have indicated that we can model the liquid either as a polymer or as a close-packed array of oxygens, atoms or ions, with a wide spectrum of cavities to take cations. In a general way, liquids have structures related to the solids that form them, except that long-range order is not present and many more irregular sites are present.

In the melt formed from pure Mg_2SiO_4 or Ni_2SiO_4 the octahedral coordination of cations in the solid is more or less preserved, as shown by studies of absorption spectra. But in liquids more rich in silica and alkali metals, more

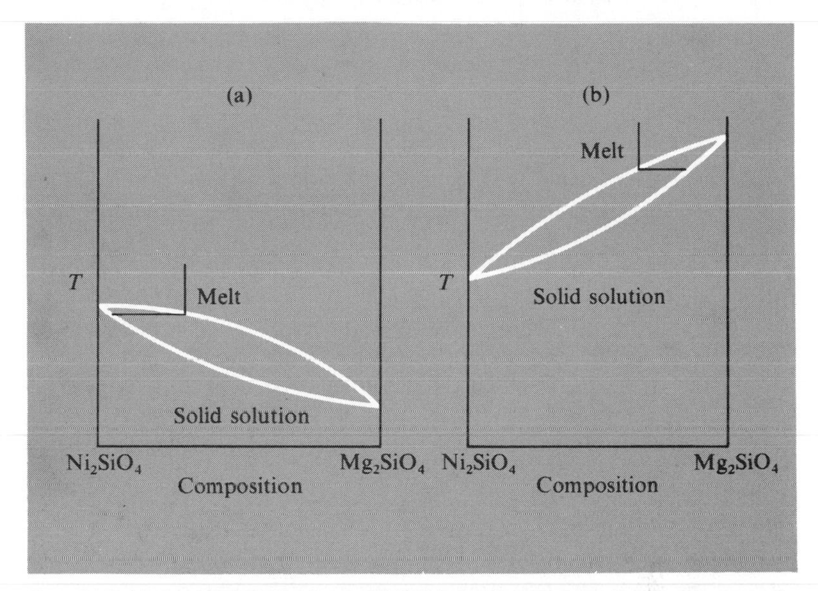

FIG. 6.16. (a) Apparent behaviour of Ni–Mg in the complex basalt system, early crystals are enriched in nickel. (b) Behaviour in the simple system Ni_2SiO_4–Mg_2SiO_4 (at much higher melting temperatures) where early crystals are enriched in magnesium.

tetrahedral and larger coordination sites are introduced. In a basaltic liquid, the number of octahedral sites is fewer and that of tetrahedral sites larger than for the pure olivine melt. Crystal-field theory tells us that the Ni^{2+} ion has a strong affinity for octahedral sites, an influence which does not occur with Mg^{2+}. Thus if we put Mg^{2+} and Ni^{2+} into a multisite liquid, Mg^{2+} will spread over the sites more than Ni^{2+}. This will lead to a larger entropy of mixing for magnesium, and if nickel is forced into octahedral sites, this will cause an unfavourable heat of mixing for nickel. The two effects mean that nickel will tend to dissolve less than magnesium in this type of liquid. Thus, liquid structure plays a large part in determining how the elements behave, and this structure changes from one magma type to another. Our knowledge of the structure of most silicate liquids is still meagre.

The phase rule

In discussing equilibria in polycomponent systems, the phase rule of Gibbs is of great value. For example if we see a rock containing phases X, Y, Z, ... we may ask the question, 'are these phases, or could these phases be, in equilibrium?'. A way of approaching this problem was introduced by Gibbs in his famous phase rule. This rule, derived from consideration of chemical potentials in an equilibrium system, is normally expressed in a form such as $F = C+2-P$, where F is the number of degrees of freedom of the system (the number of variables which must be given to define the system), C the number of components (it normally works to call this number of formula units needed to describe the composition of all phases, or, $C = N-R$ where N is the number of chemical species and R the independent relations between them), and P is the number of phases in the system. For example, in Fig. 6.12, drawn for a fixed pressure (so that $F = C+1-P$), at point X_2 three phases exist in the two-component system so that $F = 0$ and hence T must be fixed. The same is true for point E.

The phase rule distinguishes possible equilibrium states (F positive) from obviously disequilibrium* states (F negative). Many books indicate that it can be used to predict the number of phases in a system of given composition but a careful examination of the derivation will indicate that this is not true. The number of components and phases in a given system can change drastically in P–T space.

Kinetics

Although time is available in large quantities in geological systems, equilibrium states are not always achieved. Almost all rocks seen at the surface, with the exceptions of some sediments, are out of equilibrium with the surface environment. And even sediments contain a massive contribution from disequilibrium materials, aragonite rather than calcite, amorphous oxides and the like.

In all these heterogeneous systems involving solids, a number of rate processes may control what we observe; such processes often include: nucleation rates of crystals; growth rates of crystals; diffusion rates to sites of reaction; or solution rates of reactants. Homogeneous kinetic processes are not common except perhaps for some aqueous environments; and heterogeneous kinetics is a very complex subject.

Absolute reaction-rate theory expresses the rate constant of a reaction by a relation:

$$k = \frac{kT}{h}\left\{\exp\left(\frac{\Delta S^*}{R}\right)\exp\left(-\frac{\Delta E^*}{RT}\right)\exp\left(-\frac{P\,\Delta V^*}{RT}\right)\right\}$$

where ΔS^*, ΔE^* and ΔV^* refer to the entropy, energy, and volume change in forming an activated state, which may react to form products. This equation indicates that processes with a positive ΔS^* are likely to be faster than others. There is ample evidence for this geologically. Reactions which lead to disorder (melting, vaporization, etc.), tend to be more rapid than those leading to order. Of the rate processes listed above, rates of solution tend to be much faster than rates of crystal growth and in many mineralogical systems, the slow step of a transformation is the rate of nucleation.

Reaction rates naturally are more favourable at high temperatures and in a general way equilibrium tends to be approached more closely as we move away from the surface environment. More rate-controlled products occur in sediments than in igneous rocks.

On p. 33 we mentioned that when surface rocks are buried, they respond to the increasing P and T by undergoing phase changes and reactions such as dehydration. When they return to the surface, the reactions tend not to be reversed. In the metamorphic environment reactions can proceed either *via* aqueous films, the water being generated by the burial, or *via* solid-state mechanisms in crystals. Studies of diffusion coefficients in silicates show that diffusion is a very slow process and even at 1000°C, these coefficients have values in the range 10^{-12}–10^{-16} $m^2\,s^{-1}$. A very useful relation in discussing such values, is that the mean diffusion distance $\bar{X}$ of a particle in unit concentration gradient is given by:

$$\bar{X}^2 = 2Dt$$

where D is the diffusion coefficient and t the time. If D is 10^{-16}, over a year (3×10^7 s), the mean distance of motion will be about 10^{-5} m. It is for this reason that chemically-zoned crystals where the chemical potential μ of any component is continuously variable, survive through geological time.

A knowledge of diffusion coefficients is quite essential to appreciate the usefulness of minerals as memory units where element or isotope fractionation is involved and in some age-dating techniques. For example, the potassium–argon dating method depends on the assumption that when a ^{40}K atom in a crystal becomes ^{40}Ar, the product Ar will not diffuse out of the crystal. This statement is true for most crystals at low temperatures; but each specific potassium mineral has a temperature cut-off when the age memory fails, and this is diffusion controlled.

In most rock systems, solid-state diffusion processes are rarely fast enough to cause equilibrium states to be attained. In the order–disorder processes in minerals like albite ($NaAlSi_3O_8$, p. 50) it is possible to disorder a crystal by heating, but quite impossible to order it by cooling in any laboratory, or even natural, time. The feldspar exsolution process described on p. 66 rarely achieves more than an approach to equilibrium during the natural cooling of igneous rocks.

Most natural processes attain equilibrium states only if a solvent is present. For igneous rocks the solvent is a silicate melt; for metamorphic and sedimentary rocks, an aqueous fluid. Remove the fluid and reactions virtually cease and for this reason, when rocks undergoing metamorphism cease dehydrating, they become kinetically dead, and will preserve the solid phase assemblage formed during the final period of time when an aqueous solvent was present. There are some fast solid–solid reactions (e.g. the α–β quartz transition, essentially a vibrational transition) which always attain equilibrium, but for most processes involving nucleation and growth, the above holds.

In natural systems nucleation is probably rarely a homogeneous process. Surfaces abound, each surface with its own dislocation patterns, and most nucleation appears to occur on existing surfaces. Growth rates depend on dislocation densities and these can be multiplied by stress so that processes may occur faster in rocks undergoing deformation. Foreign atoms or ions tend to be adsorbed on dislocations and for this reason their presence may inhibit or promote the growth of specific phases. Thus when calcium carbonate solutions are evaporated, if the solution is pure, calcite, the thermodynamically stable polymorph, normally results. If magnesium ions are present in the solution, the high-pressure form aragonite forms. This is caused by preferential adsorption of magnesium on growth dislocations of calcite, a process which blocks further fast growth. In sediments and biological systems, such influences must be innumerable. We shall find further examples of rate-controlled phenomena when we consider the ocean–atmosphere system (Chapter 8).

7. High-temperature solutions and transport

HOT water moves through parts of the earth's crust in enormous volumes and the motion is associated with chemical transport in solution. Deposition from these solutions in mineral vein systems leads to concentration of many elements and compounds important in modern industry. Our chemical knowledge of such processes is still almost trivial.

At the outset let us consider the scale of these phenomena. Near the modern volcanic ocean ridges, melts are intruded beneath the rubbly volcanic ocean floor surface at rates of about 6 km^3 per year. The depth of intrusion is only a few km and at this depth porosity and permeability of rocks to water is large. The melts are introduced at around 1200°C. If we place a red-hot poker into a bed of wet sand, overlain by water, the mechanism of cooling will involve heat transfer *via* a convecting water system. We see this type of process in operation in many geothermal or hot-spring regions on earth. In some places where marine water is involved the resultant gases are rich in HCl owing to pyrohydrolysis of sodium chloride. In some volcanic regions, vast quantities of acid gases are liberated. For example, in the Valley of Ten Thousand Smokes in Alaska, it has been estimated that, during an eruptive period, in a single year more than a million tons of HCl was evolved, and a quarter of a million tons of HF. But as most volcanism is submarine, we do not observe the major events.

A km^3 of basalt magma will liberate about 2×10^{18} J by the time it cools to 600°C. That amount of heat could raise the temperature of 1 km^3 of sea water to 600°C. This water flows through the surface rubble and when heated will flow back to the surface. An enormous leaching or stripping experiment can be envisaged and at present, we think that many elements like iron, manganese, nickel, copper, lead, zinc, silver, and gold are concentrated by such processes.

When metamorphism affects sediments, they release water. In active tectonic regions, for example near an ocean trench, surface sediments are buried and progressively dehydrate, and the water of hydration flows back to the surface. For a km^2 crustal section of 30 km thickness (quite normal for mountain building regions), this crust section will loose about 3 per cent by weight of water or about 3 km^3 of water. The average temperature at which the water is generated is about 400°C and it flows to the surface only when the vapour pressure is about equal to the load pressure of the covering rocks (which increases by about 300 bars per km burial). Thus during metamorphism, hot and highly compressed water vapour flows from depth in vast quantities. We would like to understand the chemistry of such fluids.

Most of our ideas about inorganic solutions originate in the 25°C laboratory. There, most inorganic substances interact with water by forming hydrated ions, or some type of complex ion. Solubilities of inorganic compounds are normally discussed in terms of equilibrium constants such as ionic solubility products. If we wish to consider what may happen at elevated temperatures, we must be concerned with the entropies of these processes. Let us examine a few examples:

	Reaction	$\Delta S/(\mathrm{J\,K^{-1}\,mol^{-1}})$
$LiCl_{solid}$	$\rightarrow Li^+_{aq} + Cl^-_{aq}$	+14·2
$NaCl_{solid}$	$\rightarrow Na^+_{aq} + Cl^-_{aq}$	+43·1
HCl_{gas}	$\rightarrow H^+_{aq} + Cl^-_{aq}$	−131·4
HF_{aq}	$\rightarrow H^+_{aq} + F^-_{aq}$	−118·4
$Ca_3(PO_4)_2$	$\rightarrow 3Ca^{2+}_{aq} + 2PO_4^{3-}(aq)$	−836·8
$HSO_4^-(aq)$	$\rightarrow H^+_{aq} + SO_4^{2-}(aq)$	−109·6
$H_2S(aq)$	$\rightarrow 2H^+_{aq} + S^=_{aq}$	−567·4
$H_2CO_3(aq)$	$\rightarrow 2H^+_{aq} + CO_3^{2-}(aq)$	−244·3
$Fe^{3+}_{aq} + Cl^-_{aq}$	$\rightarrow FeCl^{2+}_{aq}$	+146·0
$Hg^{2+}_{aq} + 4Br^-_{aq}$	$\rightarrow HgBr_4^{2-}(aq)$	+51·0

A clear pattern emerges from these data. When ionic dissociation occurs, water molecules are bound to the ions to compensate for the energy of bond breaking. In general, the number of water molecules is such, that the overall loss of entropy (freedom) by the solvent molecules is greater than the gain in translational entropy of the ions. Any process leading to charged species tends to be associated with a negative entropy change and any process where total charge is reduced, (e.g. the final pair of reactions above) tends to be associated with a positive entropy change. (These ideas are discussed by G. Pass in *Ions in solution* (3): *inorganic properties* (OCS 7).) A conclusion is apparent; as elevated temperatures must favour high entropy states, high-temperature inorganic chemistry may become increasingly molecular.

There is another principle that is probably important. If we examine the vapour of molecules in equilibrium with solids at low temperatures, the vapour pressures are normally very low and the molecular chemistry rather simple. As T increases, the vapour pressure rises rapidly (cf. Fig. 6.6) and as the concentration of vapour species increases, the chance of having appreciable concentrations of associated molecules also increases. Thus with increasing T, there is a tendency for the vapour phase in equilibrium with a solid to become more complex. We thus begin to think about inorganic molecular systems with the molecular complexity rising with T. This also means that the tables of thermodynamic properties of aqueous ions at 25°C may be of little help to us, and we would like new, equally extensive, tables for inorganic molecules in solution.

Under common geothermal gradients of about 20°C km^{-1}, and pressure gradients of 300 bars km^{-1}, water remains in a fairly high-density state. For example, with these constraints, the specific volume will be about 1·0 (cm^3 g^{-1})

at the surface, 0·99 at 10 km, 1·06 at 20 km, and at 30 km and 600°C about 1·06. Over this range of conditions, the self-dissociation of water into hydrogen and oxygen increases substantially and the ionization (K_w) also changes. In fact K_w will steadily increase from the value of 10^{-14} to about 10^{-8} at 600°C and 9 kbar. Under the same conditions, the dielectric constant (relative permittivity) of water would be reduced from 80 to about 30. Thus the physical state of water does not undergo very serious changes unless the temperature gradient is steep and a low-density gas phase is present. It is only at low pressures that equilibria reflect the critical phenomena in water.

Let us now examine a typical electrolyte like NaCl in water. This is a particularly important reaction because most crustal fluids contain halogen, again a reflection of the burial of rocks with marine water in their pore system.

Sodium chloride is highly soluble in water at all temperatures, the solubility increasing steadily to the melting point. The conductivity of salt solutions has been studied over a wide range of conditions and from the conductance, the dissociation constants for the process:

$$NaCl_{aq} \rightarrow Na^+_{aq} + Cl^-_{aq}$$

determined. Values in liquid water up to the critical temperature are shown in Fig. 7.1. It will be noted that by the time the critical temperature is reached (at the critical pressure) NaCl has become a weak electrolyte with molecules predominating. The same is true for HCl (Fig. 7.1). Other values under more extreme conditions for KCl and HCl are given in Table 7.1. Simple hydroxides (e.g. NaOH, KOH) behave in similar ways. While more data are urgently needed, the trends are as one would predict.

It perhaps should be noted that if inorganic solids give molecular solutions at high temperatures, then the thermodynamic treatment of such solutions become more analogous to that for simple gas mixtures. It has been noted, for example, that near 600°C dilute mixtures of sodium hydroxide molecules in water behave in an almost ideal manner.

The molecular dominance of low-pressure, high-temperature systems leads to interesting hydrolysis reactions. For example the reaction:

$$\underset{\text{solid}}{2NaCl} + \underset{\text{gas}}{H_2O} + \underset{\text{solid}}{SiO_2} \rightarrow \underset{\text{solid}}{Na_2SiO_3} + \underset{\text{gas}}{2HCl}$$

goes to completion at 600–700°C at low pressures. Other halogen salts behave in the same manner and account for the presence of such gases in volcanic centres.

The solubility of most minerals increases in high-pressure water. The system SiO_2–H_2O provides an excellent example. The solubility curves are shown in Figs 7.2 and 7.3. In liquid water at its equilibrium vapour pressure the solubility at first increases steadily and then falls to near zero at the critical temperature. In high-pressure water, the solubility increases steadily until

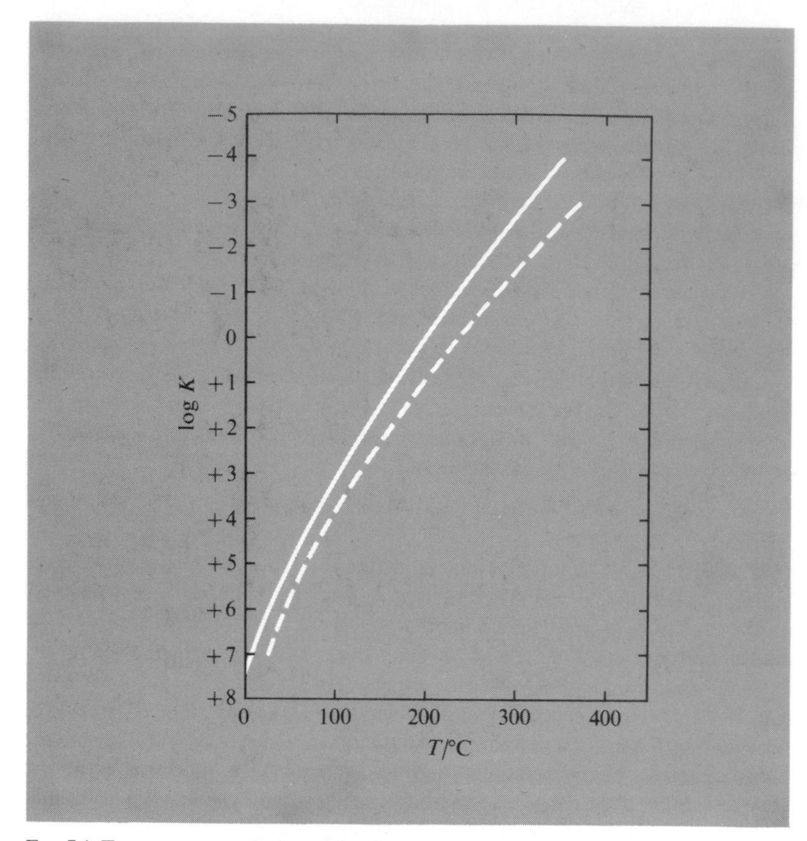

FIG. 7.1. Temperature variation of the dissociation of HCl (solid line) and KCl (dashed line) in liquid water up to the critical temperature.

melting occurs. It will be noted that under average crustal conditions (say 5 kb and 450°C) quartz solutions in water are near 0·2 molar but under extreme crustal conditions before melting occurs (say 800°C and 10 kb) the solubility has reached about 5 molar or 300 g dm^{-3}. Almost all this silica will be precipitated during passage to the surface. Above 1100°C and 10 kb, water–quartz solutions and melts become indistinguishable, the upper critical point in the system.

Most common transition-metal sulphides are highly insoluble at low temperatures. But it is known that they can be transported in large quantities at moderate temperatures. At low temperatures most sulphides form stable complexes with HS^- or S^{2-} if excess sulphide is present and the solutions are alkaline. Thus HgS is one of the most sparingly soluble sulphides but is readily

TABLE 7.1

Ionization constants at elevated temperatures

(The figures in brackets give the pressure in bars for each density of water)

(a) log K_{HCl}

T°C	Density (g cm^{-3})			
	0·3	0·5	0·7	0·8
400	−4·88	−3·66	−2·0	−0·60
	(290)	(380)	(1050)	(2100)
500	−5·05	−3·89	−2·39	−1·47
	(550)	(910)	(2040)	(3600)
600	−5·22	−4·24	−2·85	
	(810)	(1430)	(4500)	
700	−5·64	−4·55		
	(1060)	(1970)		
(b) log K_{KCl}				
450	−3·92	−2·26	−1·24	
	(410)	(640)	(1540)	
550	−4·46	−2·52	−1·89	
	(680)	(1180)	(2530)	
650	−4·68	−2·70		
	(930)	(1700)		
750	−4·91			
	(1200)			

soluble in alkaline sulphide solutions as the HgS_2^{2-} ion. Recent studies of natural hot waters has tended to indicate that transport in halogen systems may be a more general phenomena. Some figures for the composition (p.p.m.) of salt water found in bore holes at the Salton Sea, California are given below. The temperature at which samples were collected was 340°C.

Na, 54 000; K, 24 000; Ca, 40 000; Cl, 184 000; Fe, 2500;
Zn, 600; Pb, 85; Ag, 2·7; Mn, 2000; Cu, 6; S, 4.

Obviously, this is a complex salt solution containing practically no sulphur but very interesting amounts of the valuable metals Zn, Pb, Ag, Cu; in fact the water could be mined.

The solubilities of phases like sulphides in rocks depends on a series of processes, all dominantly molecular. If a halogen salt (NaCl–KCl) is present, in average crystal rocks which contain minerals such as micas, feldspars, quartz, equilibria of the type:

$$\underset{\text{mica}}{KAl_2AlSi_3O_{10}(OH)_2} + 2KCl + \underset{\text{quartz}}{6SiO_2}$$
$$\downarrow$$
$$\underset{\text{feldspar}}{3KAlSi_3O_8} + H_2O + 2HCl$$

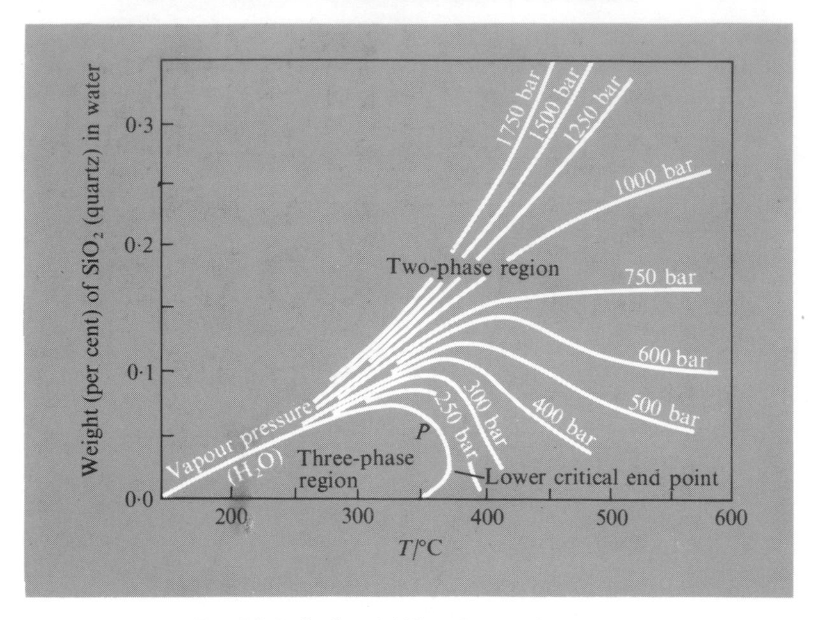

FIG. 7.2. Isobaric solubility of quartz in water.

control the partial pressure of HCl in the environment. This partial pressure rises rapidly with temperature. The HCl may react with sulphides *via* processes such as:

$$PbS + 2HCl_{mol} \rightarrow PbCl_{2,\,mol} + H_2S_{mol}$$

or more complex halogen species may form. It has been found that the solubilities of most sulphides in a system such as mica–feldspar–quartz–halide–H_2O, are quite adequate to explain ore transport.

Ore-forming process

Ore-forming processes are essentially of the same type whether they occur in the igneous, sedimentary, or metamorphic environment; a component is enriched by being removed from a volume in which it is dilute followed by transportation to, and precipitation in, a more restricted volume. Sometimes the component is a residue from the leaching and transport of less desirable components. The concentration of Al as a hydrated oxide in soils by leaching might represent the latter. The concentration of gold in quartz veins illustrates the former. We have already mentioned examples of fractionation in igneous processes (p. 67).

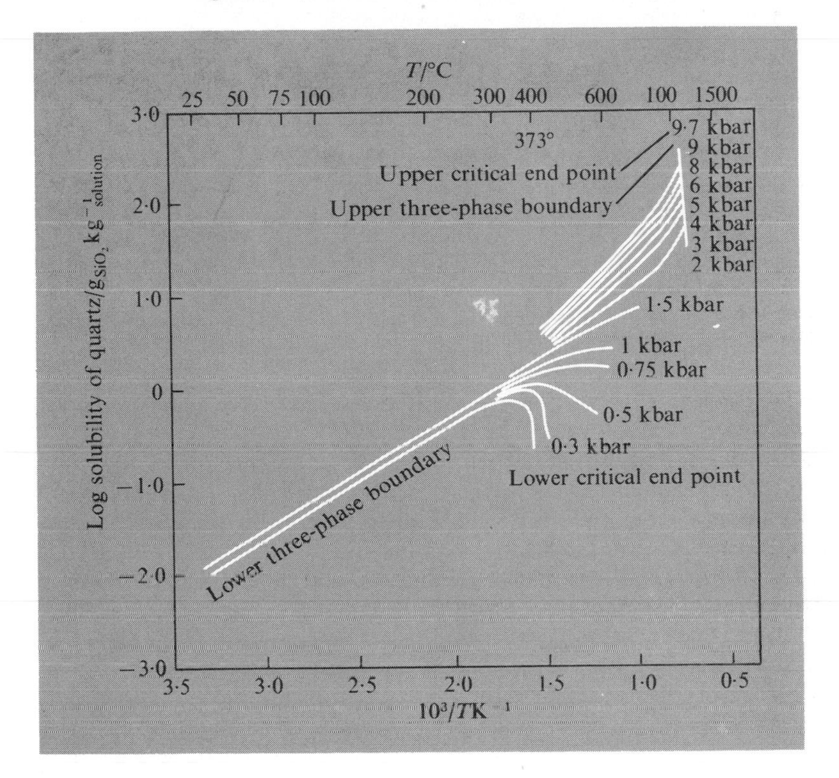

FIG. 7.3. Isobaric solubility of quartz in water up to the second critical point in the system.

During recent years, chemists have become interested in gas-phase transport in thermal gradients as a method of synthesis and purification of inorganic materials. The typical situation is indicated in Fig. 7.4. In fact, this situation is quite similar to the normal one where ores are formed (Fig. 7.5). In the simple chemical case, the rate of transport is a function of a number of terms including: the difference in concentration of the mobile component at source and sink; the path dimensions and the diffusion coefficients; the temperature or thermal gradient; the time; and changes in molecular states along the path. It should be noted that when a component i is thermally transported from a mixture $i, j, k \ldots$, enrichment or purification must occur.

In the geological situation it should be possible to quantitatively solve such transport equations. We know, or can find out something about, the thermal regime; we can set some limits on the diffusion process; in general we do not

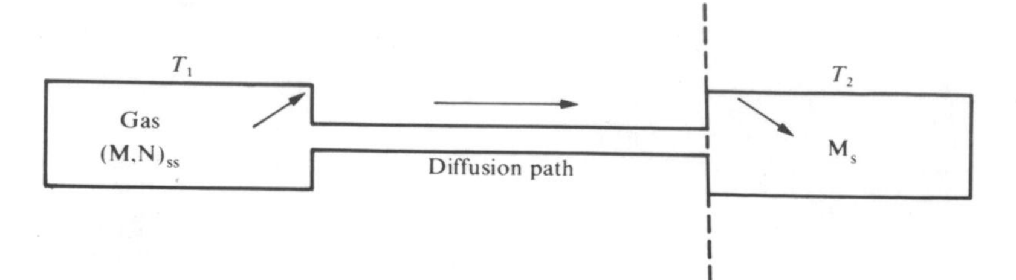

FIG. 7.4. Schematic representation of a simple transport system where a material M is purified by gas phase transport in a thermal gradient. $(M, N)_{ss}$ represents a solid solution.

know sufficient chemistry or have sufficient solution data to quantitatively estimate the transport mechanism.

Gold migration provides a good example of the type of chemistry involved. In crustal rocks gold occurs at average concentrations of 1–3 parts per billion. At this time, it can be economically extracted from rocks where it is enriched by a factor of about 10^4. Primary gold deposits normally occur when gold is deposited in quartz veins normally along with some sulphides, at temperatures in the range 300–400°C.

While gold can be inert and forms excellent containers for high-temperature aqueous experiments, it is quite soluble in oxygenated halogen-acid systems. It seems that gold in the Au^{3+} state is most easily transported. In geological systems, the oxygen regime is controlled by equilibria with the common iron minerals (p. 58). Studies have been made of the solubility of gold in a system where P_{O_2} is controlled by the reaction:

$$3Fe_2O_3 \rightarrow 2Fe_3O_4 + \tfrac{1}{2}O_2$$

and P_{HCl} is controlled by the mica–feldspar–quartz buffer system mentioned above. These studies show that in the T range 400–600°C gold dissolves *via* reactions of the type:

$$Au + \tfrac{3}{2}O_2 + 6HCl \rightarrow H_3AuCl_6 + 3H_2O$$

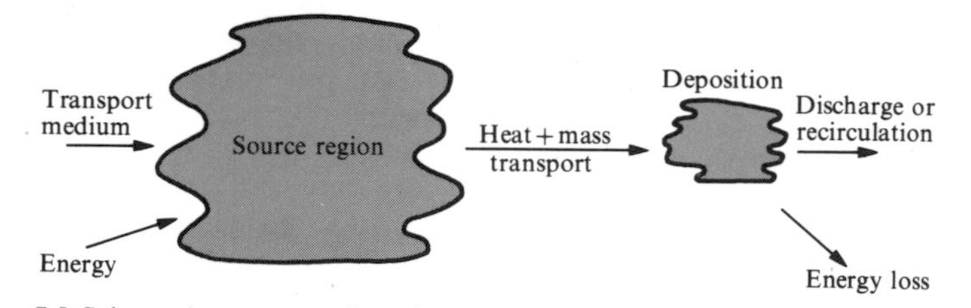

FIG. 7.5. Schematic representation of an ore forming process leading to concentration.

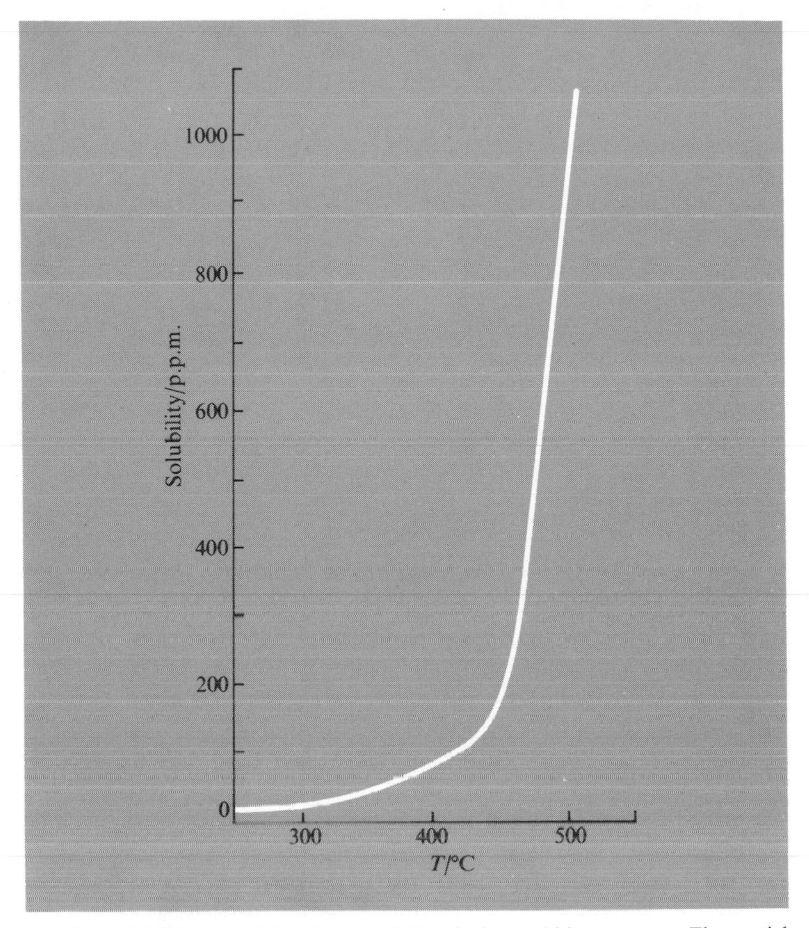

FIG. 7.6. Solubility of gold in a 2 molar KCl solution at 2 kbar pressure. The partial pressure of HCl is controlled by a buffer action with mica–feldspar (p. 77) and oxygen pressure controlled by Fe_2O_3–Fe_3O_4.

The equilibrium solubility depends on $(P_{O_2})^{\frac{3}{2}}$ and $(P_{HCl})^6$. In rocks both P_{O_2} and P_{HCl} increase sharply with temperature. It is obvious that gold solubility will also change very strongly with temperature (Fig. 7.6). A model of gold deposition based on this type of reaction appears to fit the conditions of formation of many major gold deposits. The model is shown in Fig. 7.7.

If we understood more about the chemistry of such reactions we could greatly reduce the time and expense of mineral exploration. Further, as obvious ore bodies are exhausted we must search in a more intelligent manner; the

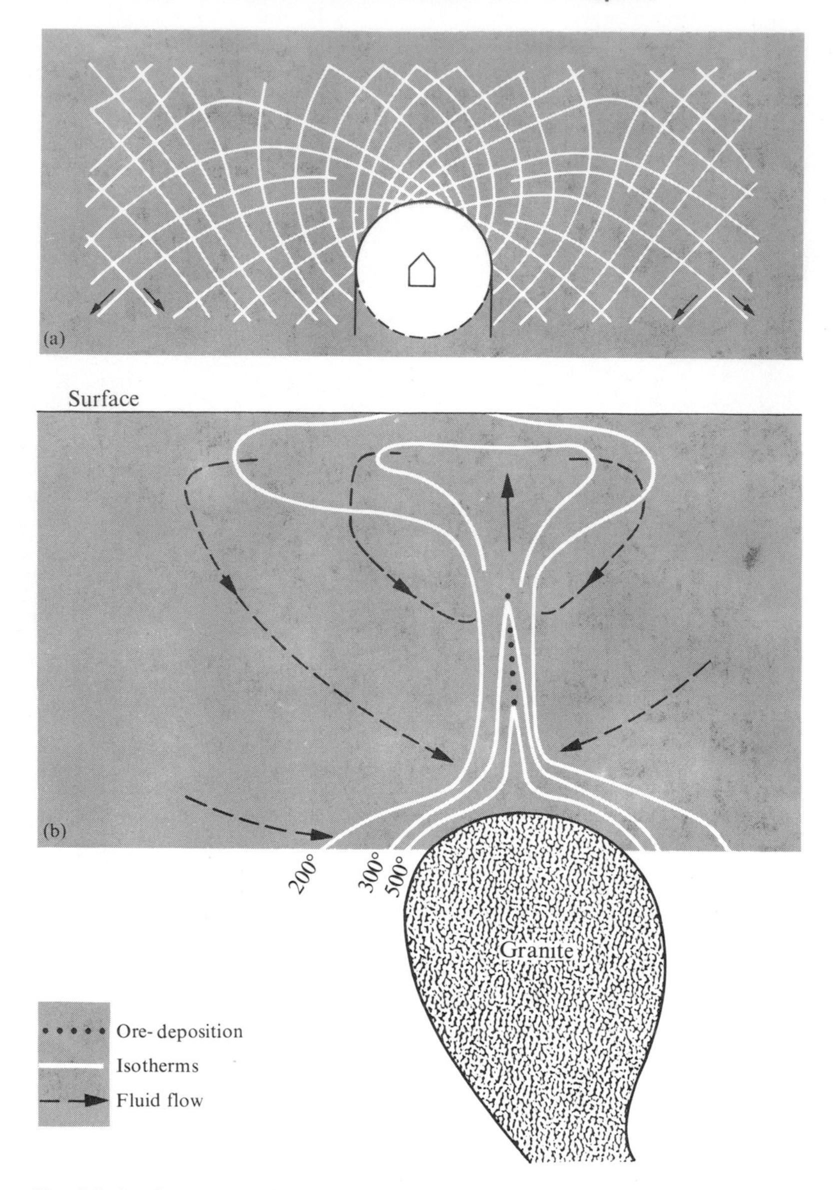

FIG. 7.7. A schematic model of a typical situation for ore deposition. A hot granite intrusion invades the crust near the surface. It causes a fracture system (7.7a) and drives a convecting flow system involving surface water. The ore material is dissolved on the heating cycle and deposited on the return cooling cycle.

earth's surface is large and ore bodies are often remarkably small. We certainly need data on the aqueous chemistry of transition-metal chloride systems to near 1000°C and fluid pressures of 10 kb.

In the search for ore bodies geochemical methods are becoming increasingly important. Near an ore body it seems logical to guess that the concentration of the ore-forming elements may be higher in percolating ground waters, soils, stream sediments, etc. This is often found to be the case and ore bodies may have a geochemical halo indicating their presence. Plants growing near ore bodies are also often enriched in a particular element so that the analysis of plant ash may be used as a prospecting tool. Today air-borne methods are becoming more useful because large areas can be sampled rapidly. Analysis of air-borne dust (soil) or infra-red 'sniffing' for elements like mercury have been used. Such methods coupled to airborne study of geophysical parameters, (radioactivity, gravity, electrical, and magnetic properties of the surface) have proved to be of immense use in prospecting over large land areas. Such methods have not yet revealed their full potential for until we really understand ore-forming processes and all the details of weathering processes and soil formation on mineral deposits, we certainly cannot appreciate the best methods of searching for ores.

8. The atmosphere and hydrosphere

THE gas–liquid–solid interface at the earth's surface is of vital importance to man. It is a complex system with interdependent rate and equilibrium processes embracing concepts from most of physical science. We are thus concerned with radiation chemistry in our magnetic field, gas kinetics, electrolyte chemistry, biochemistry, solar physics and so forth. Here brief comment only is possible on some leading geochemical observations. We are much concerned with understanding the factors that control the chemical balance of these outer regions of the earth. While lifeforms in general can tolerate amazing variation in the environment, most specific species are sensitive to minor change.

The atmosphere

The atmosphere is chemically simple and its normal composition, which remains fairly constant to heights of about 100 km, is shown in Table 8.1. One could add many other minor species, SO_2, CO, and the like, mainly associated with contributions from human activities or local volcanic events, etc.

The density of the atmosphere rapidly decreases with altitude. Near the surface the density is about 1·200 kg per cubic metre; at 100 km less than 10^{-6} kg m^{-3} and at 300 km less than 10^{-10} kg m^{-3}. At 300 km, the mean free path of atoms is about a kilometer.

TABLE 8.1

Composition of dry air (near sea level)

Species	Per cent by volume
N_2	78·08
O_2	20·95
Ar	0·93
CO_2	0·031
Ne	0·0018
He	0·000 52
Kr	0·000 11
Xe	0·000 008 7
H_2	0·000 05
CH_4	0·0002
NO	0·000 05
O_3	0·000 007 summer
	0·000 002 winter

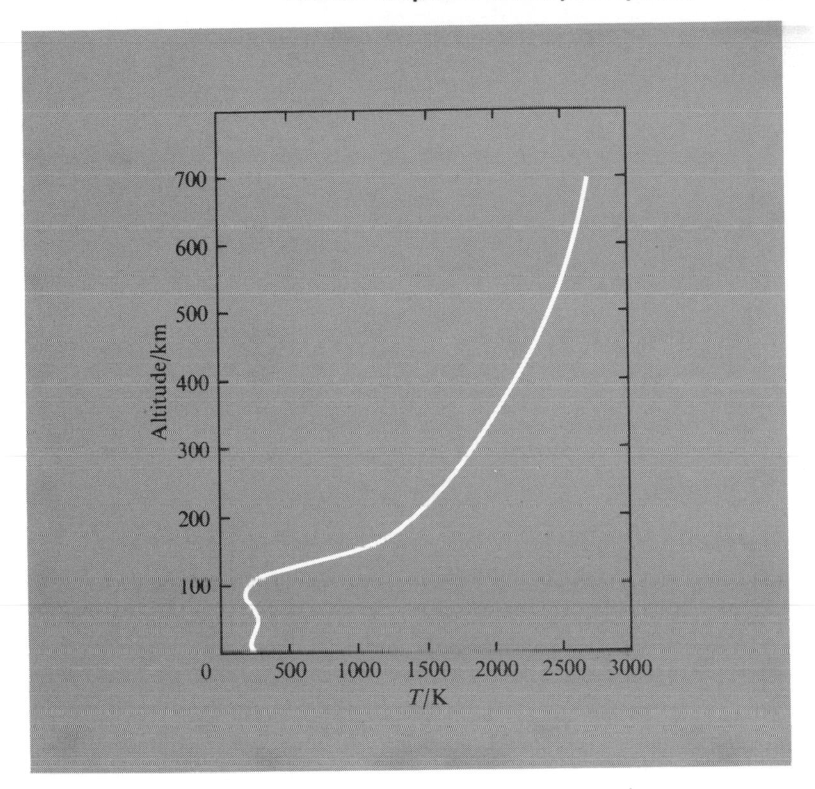

FIG. 8.1. A temperature profile for the upper atmosphere.

At low atmospheric levels the temperature profile is determined by infrared absorption in the atmosphere (mainly by water vapour) and by surface heating. Such heating falls off with elevation. At still higher levels, absorption of solar ultraviolet light and the associated photochemical reactions leads to hotter energetic gas molecules so that the temperature rises even though the gas density is now very low indeed. A temperature profile is shown in Fig. 8.1. Observed high altitude temperatures near 200 km vary from about 1000–3000 K.

Photodissociation at high levels leads to a net decrease in the mean molecular weight of the atmosphere. This is shown in Figs 8.2 and 8.3. The atmosphere also shows gravitational separation at high altitude with light O being enriched upwards. At very high levels, H and He dominate the atmosphere. By the time 200 km or so is reached, ions dominate the almost perfect vacuum. Ozone at moderate altitudes is one of the most efficient absorbers of ultraviolet

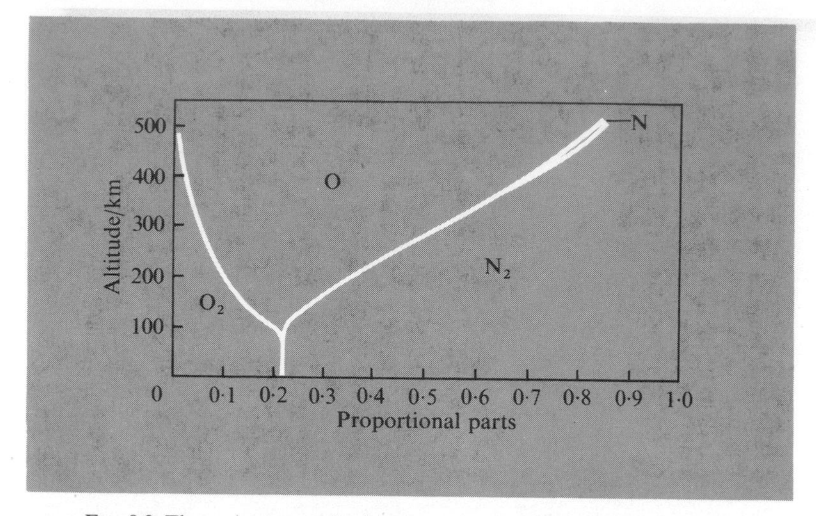

FIG. 8.2. The major atmospheric components as a function of altitude.

light which otherwise would largely eliminate life except in deep water. Complex photon absorption processes occur including:

dissociation	$O_2 + h\nu \rightarrow 2O$
ionization	$O_2 + h\nu \rightarrow O_2^+ + e^-$
excitation	$O_2 + h\nu \rightarrow O_2^*$

or all combinations of such processes. Complex recombination reactions also occur:

$$O_2^- + O \rightarrow O_3 + e^-$$
$$O^- + O_3 \rightarrow O_2 + O_2 + e^- \text{ etc.}$$

Atoms or molecules at high altitude may reach sufficient velocity to escape from the earth's gravitational field. The escape velocity for the earth

$$V_e = (2GM/R_2)^{\frac{1}{2}}$$

is about 11 km s^{-1}. The mean thermal velocity of gas molecules is given by $(3RT/m)^{\frac{1}{2}}$ but of course, molecules have a wide range of velocities above and below the mean. Only hydrogen is light enough for a significant fraction of atoms to attain the escape velocity. It has been estimated that at 2000 K (a not unreasonable high-altitude T) hydrogen concentration would fall to about one-third its value in about 1000 years. The life-time of hydrogen in our atmosphere is short and the same must be true for all hydrogen compounds for these would be photodissociated into atoms followed by hydrogen loss. The only other atom that is likely to be significantly lost in geologic time is

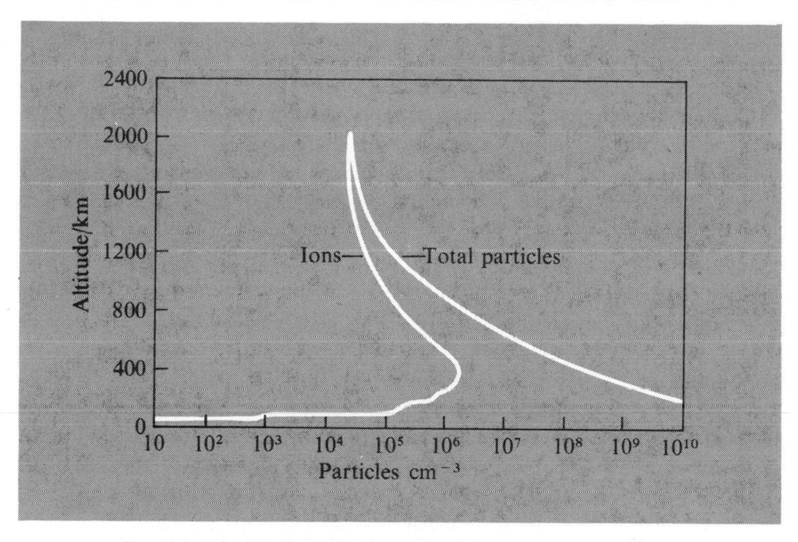

FIG. 8.3. Distribution of ions and particles in the atmosphere.

helium. Oxygen loss would involve times greater than 10^{20} years. It should perhaps be noted that the idea of organic evolution based on a dense methane–ammonia atmosphere (as found on Jupiter, 316 times more massive than the earth) is most unlikely. These gases would be very transient during the accumulation of a planet the size of the earth. But very recent observations on Titan, one of Saturn's moons, may indicate that significant amounts of these gases may be trapped in the interiors of rather small bodies. We have much to learn.

Hydrogen escape must lead to oxygen production from water and this process is one by which a slightly oxygenated atmosphere may result. The process is self-regulating. Water vapour is concentrated mainly at low levels in the atmosphere. The oxygen and ozone developed at higher levels shield the lower levels from ultraviolet radiation and thus a balance attains. Berkner and Marshall have estimated that this process would lead to a primitive atmosphere with about 0·1 per cent of the present oxygen concentration; the present high level is a result of photosynthesis in living matter.

The oxygen balance of the earth is complex. Oxygen is generated by photo-dissociation of water and by photosynthesis. It is eliminated by biological processes (decay and respiration) and by reaction with surface rocks, mainly fresh volcanics which are dominated by ferrous iron and will tend to form ferric iron compounds. Some may be lost by combination with hydrogen, carbon monoxide, and sulphur dioxide, in volcanic gases. If more oxygen were produced by photosynthesis, more carbon dioxide would be fixed in organic matter. This would lead to a reduction in the surface temperature because carbon dioxide in the atmosphere reabsorbs infrared radiation going from the surface back to space. The reduction in temperature (an ice age) would lead to diminished biological activity. Other controls are possible. Oxidation processes occur via steps involving species like O_2, $\dot{O}_3$, OH (radical) etc. If oxygen increases, then rates of oxidation will also increase being proportional to $(P_{O_2})^{\frac{1}{2}}$ for OH, O, mechanisms and $(P_{O_2})^{\frac{3}{2}}$ for ozone mechanisms. As there is an infinite sink for oxygen in the ferrous minerals of the upper mantle, and as fresh surface is generated by volcanism on a vast scale, this rate could also regulate oxygen production. The feature to be stressed here, is that our existence is regulated by such complex and vast-scale rate processes. And we are warned that a major upset in the balance of oxygen-producing organisms caused by pollution would have serious consequences. In this connection it should be noted that 90 per cent of photosynthetic oxygen is produced by organisms in the oceans and that forest land is much more productive than grass land. The atmospheric reservoir of oxygen is produced in about 3000 years of photosynthesis. For further discussion see D. J. Spedding: *Air pollution* (OCS).

Nitrogen is an inert element and almost all common nitrogen compounds are thermodynamically unstable relative to gaseous N_2. The presence of phases such as TiN and Si_2N_2O in meteorites indicates that when planetary matter accumulates, some nitrogen is likely to be fixed until an advanced stage of evolution. Once nitrogen reaches the atmosphere it is likely to be retained there. Small amounts are transformed to oxides and ammonia by atmospheric radiative processes and return to the surface *via* rain. Some organisms fix nitrogen as nitrate and ammonia and make it more available in biological activity. If organic debris in sediments is buried, the nitrogen will tend to be lost during metamorphism and return to the surface. A few ammonium analogues of Na–K silicates occur in nature but are very rare compounds.

The concentrations of the rare gases (He, Ne, A, Kr, Xe) are of considerable interest. In our atmosphere, argon dominates with smaller amounts of helium and the heavy rare gases. These abundances are drastically lower than in cosmic materials and these figures suggest that the earth accumulated from small objects that had lost most of the volatile cosmic compounds. They would not be lost from our present gravitational field, with the exception of helium. Most of the argon in our atmosphere is ^{40}A which forms from ^{40}K. Presumably it has accumulated by the slow release from potassium minerals. Helium is generated by uranium–thorium radioactive decay but the amount present in the atmosphere is much less than we might expect if there was no escape into space. It should be noted, that as the atmosphere merges into interplanetary space, the chemistry is strongly influenced by the 'solar wind' dominated by protons with a density of about 10 to 20 protons per cm^3. Studies on the moon are revealing much about the nature of particles driven from the sun.

Significant components of our atmosphere are the solid aerosol particles. These range in size from what could be termed large ions (10^{-8} m radius) up to dust particles perhaps 10 μm or so in size. As we have mentioned previously, some geochemical prospecting techniques are now based on the analysis of low-level dust particles. A large range of elements have been detected in aerosols. Some elements vital to life (Cl, S) are possibly distributed from flash-evaporated marine water. The aerosol particles so formed are later washed out in rain. Thus rain water contains about 0·5 p.p.m. Cl, 0·001 p.p.m. I, 0·4 p.p.m. Na, etc. Some aerosol particles possibly form during the entry and vaporization of meteoric material.

The hydrosphere

Most of the free water on earth is in the oceans, only about 2 per cent in lakes, streams, and ground water. A large amount of water must be locked up in the hydrated phases of the crust and mantle, perhaps an almost equal amount. Water is continuously circulated on and within the earth. Buried sediments dehydrate and the water is returned to the surface by flow through the crust or by solution in silicate melts which form volcanoes. Sea water evaporates and on the land, precipitation exceeds evaporation so that ocean water is being continuously evaporated and returned by streams and ground-water flow.

Composition of natural waters

Average sea water has the composition listed in Table 8.2. It is dominated by Na, Cl, SO_4^{2-}, Mg, Ca, K, HCO_3^-. The pH of normal sea water is usually between 7·8 and 8·3 and the mild alkalinity might be expected in a saturated calcium carbonate–bicarbonate system in equilibrium with atmosphere CO_2.

Average river water has a composition as shown in Table 8.3. It is very different both in concentrations and relative concentrations of dissolved

TABLE 8.2

Major species in sea water in g kg^{-1}

Cl^-	18·98	Mg^{2+}	1·27	HCO_3^-	0·14
Na^+	10·54	Ca^{2+}	0·40	Br^-	0·06
SO_4^{2-}	2·46	K^+	0·38	H_3BO_3	0·02

materials. Thus the order of abundance of ions is $HCO_3^- > SO_4^{2-} > Cl^-$, and $Ca^{2+} > Na^+ > K^+$; almost an inverse of marine water. The pH of surface waters is highly variable. In regions of intense organic activity where P_{CO_2} in root zones may be high, the pH may reach 4, but an average is between 6 and 8. A river like the Amazon removes about 10^5 kg of material per km^2

TABLE 8.3
Major species in river water (p.p.m.)

HCO_3^-	58·4	Na^+	6·3	Ca^{2+}	15·0
SO_4^{2-}	11·2	K^+	2·3	Fe^{2+}	0·67
Cl^-	7·8	Mg^{2+}	4·1	SiO_2	13·1

of catchment area per year. About 2×10^4 kg is in solution and the rest solid particulate or colloidal debris.

If the ocean volumes stayed constant (and there is good geologic reason for believing that there has been no major variation for a billion years or so) and if rivers continued to deliver materials to the oceans, salt concentrations would gradually build up until some form of precipitation balances the input. We are much concerned with the causes of the chemical balance in the oceans.

Let us start and follow water along this path. It evaporates from the ocean surface carrying only trivial amounts of ocean salt in aerosols. It falls as rain; water saturated with carbon dioxide in equilibrium with the atmospheric partial pressure, with very minor amounts of nitric acid, etc. Rain water has a pH around 5.7. The water falls on rock or soil surfaces and water percolates through this biologically active material, normally gaining acidity from plant debris and possibly picking up metal-complexing organic materials. It will react with mineral grains in the soil, normally by incongruent solution, and as most of these mineral grains are derived from igneous materials they are subject to alteration.

The basic reactions are of the type:

$$\text{feldspar} + CO_2 + H_2O \longrightarrow \text{(Ca–K–Na) cations in solution} + \text{clay} + Si(OH)_{4,\,\text{soln}} + OH^-.$$

The resultant solutions should be mildly alkaline. Many rocks contain carbonates and these will dissolve:

$$CaCO_3 + CO_2 + H_2O \longrightarrow Ca^{2+} + 2HCO_3^-$$

and the iron–magnesium minerals (micas, pyroxenes, amphiboles, etc.) will again dissolve incongruently *via* reactions such as

$$Mg_2SiO_4 + H_2O \longrightarrow Mg(OH)_2 \downarrow + Si(OH)_{4aq}$$

$$Mg(OH)_2 \leftrightarrow Mg^{2+} + 2OH^-.$$

The type of chemistry listed in Table 8.3 is quite in accord with what we know of the incongruent solution of the common silicate minerals. The soil leaching process tends to remove preferentially certain metals and we may pass through a series of stages with time:

$$\text{igneous minerals (Na–K–Cu–Mg–Fe–Al–SiO}_2\text{)}$$
$$\downarrow$$
$$\text{clay + iron oxides (Al–Si–Fe}_2\text{O}_3\text{)}$$
$$\downarrow$$
$$\text{oxides (Al–Fe–O–H}_2\text{O)}$$

The final oxide soil is called a laterite; it may be a source of iron or aluminium, or even at times nickel, but it is not a very encouraging agricultural material. Much of the tropical world is covered with this debris and life, whether it be plant, animal, or man, which survives on laterites tends to suffer from deficiency diseases.

Because the Gibbs free-energy change of some of these mineral reactions is very large and negative, relatively unstable intermediate products such as colloids may be formed. For example the $\Delta G^{\ominus}$ of the process

$$Fe_2SiO_4 + \tfrac{1}{2}O_2 + 2H_2O \longrightarrow Fe_2O_3 \downarrow + Si(OH)_{4,\,soln}$$

is so large, -184 kJ mol^{-1}, that the Fe_2O_3 may precipitate in an amorphous form with large surface energy and the silica in solution may be at a rather high concentration. The silica concentration in streams is normally greater than that in equilibrium with quartz, the stable modification under surface conditions. (Note: the free energy of a small particle can be expressed as

$$G_r = G^{\ominus} + \frac{2\sigma V}{r}$$

where $G^{\ominus}$ is the free energy of a 'large' grain, G_r a small grain, σ is the surface tension, V the molar volume. For most minerals the equation can be written:

$$G_r = G^{\ominus} + \frac{10^{-2}\text{–}10^{-4}}{r}\ \text{cal mol}^{-1}$$

where r is in cm. Particles of average radius 10^{-8} m may have 1–10 kcal (4–40 kJ) excess free energy. Colloidal materials are a common product of rock weathering.

Most surface waters are in equilibrium with atmospheric oxygen and oxidation reduction reactions will be largely controlled by the reaction:

$$2H_2O \rightleftharpoons O_2 + 4H^+ + 4e \qquad E^{\ominus} = -1{\cdot}299\ \text{V}$$

or

$$4OH^- \rightleftharpoons O_2 + 2H_2O + 4e \qquad E^{\ominus} = -0{\cdot}401\ \text{V}$$

or for pure water (pH = 7) $E = -0{\cdot}815$ V. This couple combined with cation potentials will allow prediction of possible oxidation states. Thus the ferrous ion will be oxidized to ferric under most normal circumstances

$$Fe^{2+} \longrightarrow Fe^{3+} + e \qquad E^{\ominus} = 0{\cdot}44\ V$$

At the pH of natural systems the product is normally a hydrated hydroxide:

$$4Fe^{2+} + 8OH^{-} + O_2 + 2H_2O \longrightarrow 4Fe(OH)_3.$$

The manganese(II) ion tends to be oxidized to the tetravalent state as in MnO_2. In soil sublayers, or regions rich in carbon, bacterial action or lack of access to oxygen can cause local reducing conditions.

Eventually the surface water finds its way to the ocean and there complex equilibria redistribute the elements. Let us examine a few simple cases to see how this works. Solid particles may remain in suspension or precipitate. Suspended clay minerals and colloids either with surface charge or large ion exchange capacities, will re-equilibrate, and possibly flocculate, on entering the new electrolyte environment. Other dissolved species must eventually precipitate or be removed by some mechanism involving precipitation, adsorption, ion exchange, etc.

Silica

The average SiO_2 concentration of stream water is around 13 p.p.m. Ocean water contains 4–6 p.p.m., the surface having a concentration in the range 0·5–2 and the bottom waters 4 p.p.m. Water in equilibrium with quartz contains about 7 p.p.m. at low temperatures. As stream water contains more SiO_2, it is obvious that it must be rapidly removed from the sea. This introduces the notion of the 'residence time' of a species. The oceans contain $1{\cdot}4 \times 10^{21}$ kg of H_2O. Water flow into the oceans is about 10^9 kg s^{-1}. Thus it takes $1{\cdot}4 \times 10^{12}$ s to fill the oceans and this time, $4{\cdot}4 \times 10^4$ y, is termed the residence time of water. The residence time of any species in the ocean is defined as the number of years it takes to renew the total amount in the ocean. For silica, this time is about 0·02 million years. Silica removal is achieved by the formation of new silicate phases; clay, zeolites, and the like, and it is removed in massive quantities by radiolaria and other micro-organisms which secrete a skeleton of amorphous silica. Biological energy must be consumed in the process because the free energy (and the solubility) of amorphous silica is much higher than that of silica in sea water. Biological activity is concentrated in the surface layers with access to light and this probably accounts for the lower surface concentrations of silica. During submarine volcanic eruptions, vast quantities of silica may be pumped into sea water by hydrothermal leaching of hot rocks. Where this occurs silica-secreting organisms may bloom, and rocks (cherts) may be precipitated by the accumulation of their skeletal remains. Thus biological buffering appears to be important.

Manganese

In most natural waters (pH 6–8) manganese is transported in solution as manganese(II) ions although some may be present in complexes like $MnSO_4$ (aq). Again the residence time is very short, being only about 1000 years. It is well known that many parts of the ocean floor are covered with manganese precipitates or nodules. These are hydrated solids dominated by the Fe–Mn oxide phases MnO_2, MnOOH, FeOOH, etc. They often contain rather large amounts of Co, Ni, and Cu, and are valuable potential sources of all these elements. In the Pacific ocean they cover nearly 10 per cent of the ocean floor and up to 50 kg m^{-2} has been observed. They grow in size at a rate of a few millimeters per million years, but in some volcanic situations probably much faster.

The manganese(II) ion concentration in the normal ocean is about 4×10^{-8} mol dm^{-3}. From thermodynamic data of the reaction:

$$4Mn^{2+} + O_2 + 6H_2O \rightarrow 4MnOOH + 8H^+$$

and for pH = 8, $P_{O_2} = 1$, we would estimate an Mn^{2+} concentration of 3×10^{-12} mol dm^{-3} as equilibrium with the oxide. It is no surprise that a precipitate forms or that the manganese(II) concentration of the sea is very low. Certain bacteria are efficient at precipitating manganese oxides and it is even possible that the formation of oceanic nodules involves bacterial action.

While the residence times for silica and manganese are calculated from an even input, a large contribution may also come from submarine hydrothermal action. We have seen (p. 77) how hot, reduced, salt waters can dissolve manganese. Sea water convecting through hot volcanic rocks could strip large quantities of most transition metals and silica and all these elements are seen to be enriched in chert deposits associated with basaltic submarine lava flows. Many of the major Cu–Pb–Zn–Ni–Ag–Mn ore deposits may have their primary origins in such processes.

Barium

The amount of barium in ocean water is almost exactly what would be expected from the solubility of $BaSO_4$. Certain marine organisms (protozoa) precipitate $BaSO_4$ in their cells.

Chlorine

As we have seen above, stream waters normally contain only small amounts of chloride and as most common halides are soluble, the residence time of chlorine in the oceans is large, perhaps about 10^8 years. Salinity of natural waters is quite variable; for example the land-locked Dead Sea has much larger salt concentrations than the normal oceans. We also know from evaporation experiments, that higher salinities can be tolerated by rocks. What then keeps the oceans in general at a more or less constant composition, particularly with respect to halide ions?

Before we can answer this question we must know the abundance of chlorine in the earth. There are very few common minerals in the crust which contain much chlorine. Sodium chloride occurs in evaporites on a fairly large scale, salt occurs in pore waters of sediments but estimates of the total quantities show these contributions to be minor relative to the quantity present in the oceans. Common igneous rocks of the crust contain only about 100 p.p.m. chlorine. We would not expect much chlorine in the mantle because this element forms such volatile hydrides. For the same reason we would not expect large quantities of chlorine to be trapped in the primary accumulation process. All in all, it seems that the primary reason for the chlorine content of the oceans, is simply that most of the available chlorine is in fact in the oceans. If we added all the chlorine in the crust, it would only increase the ocean content by 10 per cent or so.

But chlorine is significantly mobile. The convective system of the earth returns ocean floor materials to the mantle and these rocks will have a pore system filled with sea water. As the rocks are deeply buried this water will be eliminated and returned to the surface during compression and heating.

Evaporation processes affecting marine water may also eliminate massive amounts of chloride. Such processes occur today on a large scale in the Middle East and throughout the geologic record there are massive salt deposits often associated with oil fields. When the Atlantic Ocean opened to separate South America from Africa, it is clear that the process was not steady. Vast evaporite deposits which match in West Africa and East Brazil were formed when the initial mini-Atlantic oscillated and closed and evaporated almost to total dryness. Eventually, most evaporite salt will be returned to the oceans. There are very few ancient evaporites. Thus it is mainly crustal mixing phenomena that prevent virtually all the chlorine of the earth being in the oceans. It should be noted that sediments formed during the $3{\cdot}5 \times 10^9$ years during which we have a record, are remarkably similar. There is little convincing evidence that ocean water chemistry has ever changed substantially.

Sodium

Along with chlorine, sodium has a large residence time, much longer than potassium which is preferentially precipitated in clay minerals. Some sodium is removed by the same mechanisms as indicated above for chlorine but the general balance of cations in the oceans is mainly a result of ion exchange equilibria involving the clay minerals.

Reducing conditions

Most of our discussion above has been concerned with water in equilibrium with oxygen. But below the sediment interface, sulphate-reducing bacteria may be active and sulphide formation may precipitate many insoluble sulphides. As circulation in the upper metres of a sediment is possible (or at least diffusive communication), this is a vital region where precipitation may occur.

Elements like Mg–Fe are precipitated in clay and chlorite minerals. Reactions of the type:

$$\text{Fe clay} + Mg^{2+} + SO_4^{2-} + \text{(bacteria)}$$
$$\downarrow$$
$$\text{Mg clay} + FeS_2$$

contribute to the balance of Fe, Mg, S. In the same environment elements like silver, lead, and zinc, would also be fixed as sulphides. Bacterial sulphate reduction may also be important in regions where hydrogen-bearing geothermal waters rise through sediments bringing with them appreciable concentrations of transition metals. Recent work also indicates that vast quantities of ocean sulphate may be fixed in basalts as sulphide or even sulphate by sea-water convection into the ocean floor.

Calcium carbonate

Reactions in the oceans involving calcium and dissolved carbon dioxide are significant in the calcium balance and the CO_2 balance of the atmosphere. Just as with oxygen, the CO_2 balance is vital to the thermal stability and biological productivity of the environment.

Most of the carbon in the crust is present in carbonate sediments ($CaCO_3$, $MgCO_3$) and a fairly large amount as the remains of organic materials (coal, oil, and the like). Of the freely-available carbon in carbon dioxide, about 2 per cent is in the atmosphere and the rest in the oceans. As ocean pH is controlled by silicates and carbonates (excess base) and is about 8, most of the available carbon is present as bicarbonate and carbonate ions.

It has proved a very difficult task to understand fully the nature and thermodynamic activities of the carbonate species in ocean water with its rather large ionic strength. Recent work however, has largely solved these problems. In Table 8.4 we show the major species in sea water. It will be noted from these tables, that with the major cations most are present as free ions. But the anions, and in particular sulphate and carbonate, are significantly locked up in ion-pair complexes. Activities for the major ions in sea water are given in Table 8.5. At the ionic strength of sea water, divalent complexes have low activity coefficients and are complexed. Thus while the total molality of carbonate in the sea is $0{\cdot}27 \times 10^{-3}$, the thermodynamic activity is only $0{\cdot}0057 \times 10^{-3}$. Until these data were available, it was quite impossible to answer questions regarding solution equilibria of carbonate species.

TABLE 8.4

Major ionic species in sea water

	$10^3 \times$ Molality	% free ion	% SO_4^{2-} ion pair	% HCO_3^- ion pair	% CO_3^{2-} ion pair	
Na^+	475	99	1	–	–	
K^+	10	99	1	–	–	
Mg^{2+}	54	87	11	1	0·3	
Ca^{2+}	10·4	91	8	1	0·2	
			% Ca^{2+} ion pair	% Mg^{2+} ion pair	% Na^+ ion pair	% K^+ ion pair
SO_4^{2-}	28·4	40	3	19	39	0·5
HCO_3^-	1·8	74	3	15	9	–
CO_3^{2-}	0·27	10·5	7	63·5	19	–
Cl^-	550	100	–	–	–	–

TABLE 8.5

Activities of major ions in sea water

	$m \times 10^3$	activity $(\gamma m) \times 10^3$
Cl^-	550	346
Na^+	475	332
Mg^{2+}	54	13·5
SO_4^{2-}	28·4	1·94
Ca^{2+}	10·4	2·39
K^+	10·0	6·2
HCO_3^-	1·8	0·92
CO_3^{2-}	0·27	0·0057

In a general way, it has been found that near-surface waters are about saturated with calcium carbonate. Local variation of biological activity affects P_{CO_2} and hence the equilibrium. At depths below a few hundred metres pressure effects on the solubility of calcite lead to general undersaturation. When the skeletons of calcareous organisms fall from near the surface to deep water, they tend to dissolve. The bulk of carbonate precipitation, either as stable calcite or the high-pressure metastable polymorph aragonite, occurs in biological systems.

It is interesting to contemplate what might follow if we destroyed organisms that influence the CO_2 balance by carbonate precipitation. One might expect that higher levels of supersaturation would be required for bulk inorganic precipitation. This could mean less CO_2 in the atmosphere, which could significantly change the thermal balance of the atmosphere.

While the calcium carbonate system must provide one of the buffers on the CO_2 balance other systems also operate. Such a system is:

$$\text{clay} + Na^+ + HCO_3^- \rightarrow \text{Na clay} + CO_2$$

and this will be associated with the exchange reaction:

$$\text{Ca clay} + 2Na^+ \rightarrow Na_2\,\text{clay} + Ca^{2+}$$

Thus silicate reactions are also important although the time constants of response of the two types of reactions may be rather different. Carefully controlled bio-inorganic experiments are required to elucidate such problems.

We perhaps should note that oceanic phosphate is controlled by the precipitation of apatite ($Ca_5(PO_4)_3OH$–$Ca_5(PO_4)_3F$). Most apatite is precipitated biologically and phosphate equilibria are analogous to carbonate systems.

In summary, from these examples we see the ocean–atmosphere system as being regulated by a complex array of inorganic and biochemical processes. The sediment interface, where conditions with respect to oxygen access change, is probably also vital to our detailed understanding. Add to this the additions from massive submarine volcanism and sediment return to the upper mantle *via* convective processes which form the ocean floors, and we begin to appreciate the complexities of this system.

Evaporites

When total or partial evaporation of marine and fresh water occurs, useful salt deposits are formed. From evaporites we obtain most of our supply of borates, NaCl, KCl, $CaSO_4$ (as the hydrate gypsum), etc.

If sea water is progressively evaporated precipitates occur in the order: $CaCO_3$, $CaSO_4$, NaCl, and eventually chlorides, sulphates and bromides of Na–K–Mg. Evaporation of a 100 metre ocean-water column would yield a 1·50 m salt layer of which 1·16 m would be NaCl. Total evaporation is a rare phenomena and partial evaporation in a region of restricted circulation is more usual. In such a region, water is periodically replenished and the most soluble salts are not deposited.

In some continental desert regions evaporation leads to the precipitation of phases such as sodium carbonates and borates and at times even nitrates as in the famous deposits of Chile.

When salt deposits are buried and metamorphosed, on account of the large amounts of water which are involved (p. 73) most soluble salts are eliminated and the components are returned to the surface. The only traces of the evaporite may be found in carbonate–sulphate deposits which can survive if they are massive. Records of such deposits are found in rocks over 2×10^9 years old.

9. Evolution of the earth

MOST theories of the origin of the earth and solar system involve some type of expanded protosun with a disk of cosmic materials in which periodic condensations occur to form the planets. Such models are in accord with the laws of conservation of momentum and the observed motions of the planets. As mentioned on p. 88 it is also certain that the materials lost most of the light cosmic gases before larger bodies were formed and that preliminary accumulation involved small, cold objects. It seems possible too, that there was element separation in the disc which may account for some of the non-cosmic or non-solar abundance ratios. It has even been suggested that magnetic fields may have influenced the position of paramagnetic species such as the transition elements.

There is considerable divergence of opinion as to exactly how the material of the earth accumulated. Some theories suggest that small objects formed, perhaps up to the size of asteroids, and that their composition was rather like the chondritic meteorites. Following the accumulation, the protoearth heated up by gravitational rearrangement and by radioactivity which might include activity of short-lived elements present in the newly formed array of nuclides. All theories indicate a rather short time, a few hundred million years or less, between element synthesis and the appearance of solid objects. It should be noted that energy from gravitational separation is very large indeed. If we were to form a homogenous earth, and then separate the metallic core, the energy evolved would be sufficient to melt a large fraction, perhaps 50 per cent, of the earth. During this melting process, and the separation of the core, a vast distribution experiment would occur, some elements like Au, Pt, and S, would concentrate in the core while oxyphile elements and volatile elements (U, K, and W) would rise toward the surface.

Other theories suggest that the carbonaceous chondrites might represent a better starting material. Formation of metal is associated with carbon reduction of oxides during heating up. In essence the later steps of the process are as above.

A popular theory today, is that formation of the original solid objects involved a fractionation process based on the equilibrium vapour pressure of potential compounds in a cosmic gas. Then phases like Al_2O_3 and other refractory oxides would condense first, followed later by metals, and more volatile feldspar-like silicates. In this way, the accumulated material would be partly fractionated before the entire body formed and would be a little like the present earth. But reactions in a cosmic gas cloud at low temperatures could well be rate-determined and equilibrium parameters alone may not lead to reasonable prediction.

Like so many complex problems, it could well be that the truth is a mixture of all theories. What we do know is that in a broad sense the earth shows an approach to gravitational equilibrium. There is evidence that the thermal gradient was higher in the past and that melting could have been more pervasive providing a lower viscosity earth in which separation could have been fast.

The oldest rocks on earth, now approaching 4×10^9 years, indicate that the crust was highly evolved by this time, the rocks having similar modern counterparts. The oldest sediments ($> 3{\cdot}5 \times 10^9$ yrs) are also rather normal and indicate that oceans were highly developed at that time. Volcanic rocks and submarine volcanic rocks, are common in the oldest crust while granites dominate the geological maps of preserved remnants of ancient regions. As soon as water was present on the surface, a crust about 10 km thick would be rather stable and this crust would be dominated by the low-melting and light fractions. Questions as to whether oceans were more or less voluminous remain unanswered but there is no convincing data indicating that their chemistry was significantly different (cf. p. 94).

In general, sediments show little evidence that the early atmosphere was different although this is still a subject of debate. It seems possible that, in the absence of photosynthetic organisms, the oxygen content was lower. This would mean less u.v. screening of the surface (p. 87) and while a lower concentration of oxygen would cause slower oxidation rates, a higher surface flux of u.v. radiation would cause a higher concentration of free radicals active in oxidation. Studies indicate that the two effects might about balance out.

There is also evidence of biosynthetic organic materials in the oldest rocks. Life seems almost as old as surface rocks themselves. The vast accumulations of iron oxides in older rocks may reflect enhanced volcanism bring ferrous iron to the oceans and intense bacterial precipitation. The origin of these rocks (which provide most of the iron utilized on earth today) is still unresolved. Abelson has shown that paths leading to molecules necessary for life can be devised *via* HCN produced in a CO_2–H_2O–N_2 atmosphere. Once produced, there are rather facile paths to species like amino acids. In this connection it is interesting to note that recent studies have revealed quite complex molecules in space: HCN, CH_3CHO, CH_2NH, HNCO, and the like.

The biomass has been of immense significance in earth history. The mass of the oceans is about $1{\cdot}4 \times 10^{24}$ g, the entire Earth, $5{\cdot}9 \times 10^{27}$ g. At the present time organic matter is synthesized in living cells at the rate of about 10^{17} g per year. The larger part is produced in the oceans. At the above rate, in the period of earth history (4.5×10^9 years), a mass of $4{\cdot}5 \times 10^{26}$ g could be produced. Most of the biological stuff (C, O, N, S, etc.) is recycled but the figures are impressive. Biological precipitation of inorganic materials is equally impressive. In our present age, man is furiously engaged in oxidizing the fossil remnants of past organic activity.

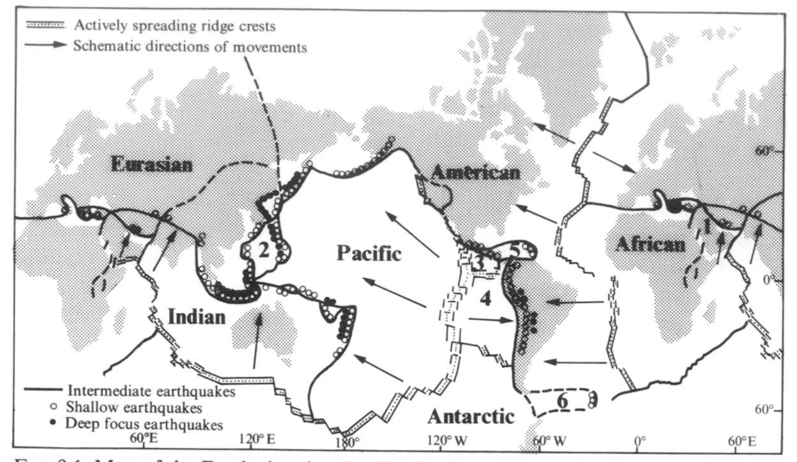

FIG. 9.1. Map of the Earth showing the distribution of the major plates, ocean ridges (where basalts appear from the mantle), subduction* zones which correspond with deep earthquakes which in turn correspond with regions of andesitic volcanism and granite plutonisms. Small plates are numbered. (1) Arabian, (2) Phillippine, (3) Cocos, (4) Nasca, (5) Caribbean, (6) Scotia. The Mediterranean region is very complex.

With time, there is evidence for a decreasing thermal gradient and a build-up in atmospheric oxygen although attainment of present levels may have occurred only during the last 4×10^8 years or so.

Superposed on equilibrium trends associated with a cooling body, we have the massive effects of convection by which deep materials arrive at the surface by volcanism and surface materials are drawn back into the upper mantle by descending convection currents. The modern earth is discussed in terms of large, rather stable surface regions termed plates (about half a dozen) which are in a state of motion relative to each other. The plates are 100–150 km thick and volcanic activity is concentrated at their margins. The volcanic activity is basaltic where new crust is formed and andesitic and granitic, where crust is dragged down or subducted (Figs 9.1 and 4.3). Plate margins can be eaten away and old materials returned to the mantle. The miracle is that in such a dynamic system, old crust is preserved at all.

Modern convection cells are large. Were they larger or smaller in the past? Again the question is not resolved. Some workers consider larger cells existed in the past, which swept the continental crust into a single mass; some consider that there is evidence for much smaller cells in the past. The problem is that much of the evidence is swept away by the modern motions. Many questions such as the distribution of the crust in the past, the volume of the oceans, etc., remain unresolved. The answer to questions about the ancient past must be found in rocks for we cannot make geophysical observations. Chemistry or geochemistry becomes even more important in studies of the ancient earth regime.

What of the future? The earth will cool, but very slowly. As it cools, more water will become locked in hydrated phases of a cooler crust and mantle. If convection does not stop, gradually the oceans will vanish. Or if convection slows down, the oceans will become rather static and the land surface removed till the entire earth is covered by water. Hydrogen will slowly bleed off into space. Or will the sun evolve first, and go through its dying phase of high luminosity when the oceans will boil off? But as Gamow states 'there is no reason for immediate panic—we still have five billion years to go'.

Bibliography and notes

GENERAL REFERENCES

GOLDSCHMIDT, V. M., *Geochemistry.* (Ed. A. Muir). Clarendon Press. 1958.
This is a classic written by the man who can be considered the founder of modern geochemistry. Goldschmidt's interests covered the entire spectrum of geochemistry even extending into agricultural chemistry. He proposed many of the generalizations concerning factors governing the distribution of the elements. He was responsible for one of the first extensive determinations of ionic radii. This book is still an important reference work in geochemistry.

CLARK, S. P., Handbook of physical constants. *Geol. Soc. Am., Memoir* 97, 1966.
A valuable source of data of geochemical and geophysical properties of rocks and minerals.

FAIRBRIDGE, R. W., *The encyclopedia of geochemistry and environmental sciences Encyclopedia of earth sciences series, vol. IV A.* Van Nostrand Reinhold Co., New York, 1972.
Encyclopedias are often disappointing. This one is not. It contains valuable short accounts of many facets of geochemistry, each written by an expert in the field. An excellent reference work.

WEDEPOHL, K. H., *Handbook of geochemistry.* Springer-Verlag, Berlin, 1969.
A most valuable work concerning the detailed geochemistry of each element with accounts of the major general principles.

ROBIE, R. A. and WALDBAUM, D. R. Thermodynamic properties of minerals and related substances at 298·15 K (25·0°C) and one atmosphere (1·013 bars) pressure and at higher temperatures. *U.S. Geological Survey Bull.* 1259, 1968.
A standard reference work for thermodynamic data of minerals.

KRAUSKOPF, K. B. *Introduction to geochemistry.* McGraw-Hill, New York, 1967.

MASON, B. *Principles of geochemistry.* John Wiley, 3rd edition, New York, 1966.
The above two books are excellent general introductions to the subject.

BURNS, R. G. *Mineralogical applications of crystal field theory.* Cambridge University press. 1970.
A general account of the application of modern concepts to the geochemistry of the transition metal elements.

BERNER, R. A. *Principles of chemical sedimentology.* McGraw-Hill, New York, 1971.
An advanced treatment of equilibria in surface waters of complex chemistry.

GARRELS, R. M. and CHRIST, C. L. *Solutions, minerals and equilibria.* Harper and Row, New York, 1965.
A basic work mainly concerned with equilibria in natural aqueous systems.

MASON, B. *Meteorites.* John Wiley, New York. 1962.
An excellent summary of the most important observations on meteorites.

VERHOOGEN, J., TURNER, F. J., WEISS, L. E., WAHRHAFTIG, C., and FYFE, W. S. *The Earth*. Holt, Rinehart and Winston, 1970.

An elementary general account of geology written for the student of physical sciences.

SKINNER, B. J. and TUREKIAN, K. K. *Man and the ocean*. Prentice-Hall, Englewood Cliffs, New Jersey, 1973.

A valuable simple account of the nature of the oceans and problems of conservation.

ROBERTSON, E. C., *The nature of the solid Earth*. McGraw-Hill Inc., New York, 1972.

This multi-author book (Robertson editor) contains an excellent series of essays on modern geophysical problems concerning the structure of the earth from crust to core.

WYLLIE, P. J. *The dynamic Earth*. John Wiley, 1971.

An excellent modern account of major dynamic problems associated with modern theories of convection and melting and phase changes in the earth.

JOURNALS

The most important geochemical journals in English are *Geochimica et Cosmochimica Acta* published by Pergamon Press and *Chemical Geology* published by Elsevier. Much of the enormous Russian effort in this field is summarized in the English translation of the Russian *Geokhimiya*; *'Geochemistry International'*. But many important studies are scattered through the literature of earth sciences and perhaps *Chemical Abstracts* is still the most useful general reference source.

RELATED BOOKS IN THE OXFORD CHEMISTRY SERIES

EARNSHAW, A. and HARRINGTON, T. J. *The chemistry of the transition elements.*
PASS, G. *Ions in solution* (3): *inorganic properties.*
PUDDEPHATT, R. J. *The periodic table of the elements.*
SMITH, E. B. *Basic chemical thermodynamics.*
SPEDDING, D. J. *Air pollution.*

Glossary

billion: 10^9.
chalcophil(e) elements: those tending to concentrate in sulphide ores.
chondrules: small rounded bodies, typical of some stony meteorites.
core: the central part of the Earth's interior.
crust: the outermost layer of the Earth, above the Moho (q.v.).
disequilibrium: non-equilibrium.
end-member mineral: one of two minerals forming a solid solution series, with continuously variable composition; e.g. Mg_2SiO_4 (forsterite) and Fe_2SiO_4 (fayalite).
exsolution: appearance of two mineral phases in the solid state on cooling to a specific temperature (the *exsolution temperature*) e.g. perthitic intergrowths in feldspars.
gabbro: plutonic igneous rock of coarse grain and basic (basaltic) composition composed essentially of plagioclase and clinopyroscene, sometimes together with olivine.
hydrosphere: collective term for the oceans and all waters on or near the earth's surface.
lithophil (e): elements: those tending to concentrate in stony (silicate) matter.
magma: molton rock-material.
magmatism: formation of igneous rocks and magma.
mantle: the part of the earth's interior between the core and the crust.
mascons: (mass concentrations) concentrations of denser material below the lunar maria postulated to account for observed gravity anomalies.
Moho: the Mohorovičič discontinuity; interface (about 35 km below continents, 12 km below oceans) at which seismic velocities increase sharply; boundary between mantle and crust.
ocean trench: deep trench-shaped region near margin of a continent and parallel to it, resulting from descending convective flow in the mantle.
paleomagnetism: rock magnetism; the study of the magnetism of rocks acquired in the geological past.
plumes: rising (hot) or descending (cold) regions of convection cells in the mantle.
pluton: an igneous rock mass crystallized at depth in the crust.
plutonic: pertaining to the action of heat at depth. Plutonic rocks are igneous rocks formed at great depth in the crust.
seismograph: instruments for recording earthquakes.
siderophil(e) elements: those tending to concentrate in metallic iron.
subduction: process by which crust is consumed by being forced down into the mantle insubduction (or Benioff) zones at plate margins.
tectonic: relating to movement and fracture of rocks.

Index

1A	IIA	IIIA	IVA	VA	VIA	VIIA	VIII			IB	IIB	IIIB	IVB	VB	VIB	VIIB	O
$_{1}$H 1·008																	$_{2}$He 4·003
$_{3}$Li 6·941	$_{4}$Be 9·012											$_{5}$B 10·81	$_{6}$C 12·01	$_{7}$N 14·01	$_{8}$O 16·00	$_{9}$F 19·00	$_{10}$Ne 20·18
$_{11}$Na 22·99	$_{12}$Mg 24·31											$_{13}$Al 26·98	$_{14}$Si 28·09	$_{15}$P 30·97	$_{16}$S 32·06	$_{17}$Cl 35·45	$_{18}$Ar 39·95
$_{19}$K 39·10	$_{20}$Ca 40·08	$_{21}$Sc 44·96	$_{22}$Ti 47·90	$_{23}$V 50·94	$_{24}$Cr 52·00	$_{25}$Mn 54·94	$_{26}$Fe 55·85	$_{27}$Co 58·93	$_{28}$Ni 58·71	$_{29}$Cu 63·55	$_{30}$Zn 65·37	$_{31}$Ga 69·72	$_{32}$Ge 72·59	$_{33}$As 74·92	$_{34}$Se 78·96	$_{35}$Br 79·90	$_{36}$Kr 83·80
$_{37}$Rb 85·47	$_{38}$Sr 87·62	$_{39}$Y 88·91	$_{40}$Zr 91·22	$_{41}$Nb 92·91	$_{42}$Mo 95·94	$_{43}$Tc 98·91	$_{44}$Ru 101·1	$_{45}$Rh 102·9	$_{46}$Pd 106·4	$_{47}$Ag 107·9	$_{48}$Cd 112·4	$_{49}$In 114·8	$_{50}$Sn 118·7	$_{51}$Sb 121·8	$_{52}$Te 127·6	$_{53}$I 126·9	$_{54}$Xe 131·3
$_{55}$Cs 132·9	$_{56}$Ba 137·3	$_{57}$La 138·9	$_{72}$Hf 178·5	$_{73}$Ta 180·9	$_{74}$W 183·9	$_{75}$Re 186·2	$_{76}$Os 190·2	$_{77}$Ir 192·2	$_{78}$Pt 195·1	$_{79}$Au 197·0	$_{80}$Hg 200·6	$_{81}$Tl 204·4	$_{82}$Pb 207·2	$_{83}$Bi 209·0	$_{84}$Po (210)	$_{85}$At (210)	$_{86}$Rn (222)
$_{87}$Fr (223)	$_{88}$Ra 226·0	$_{89}$Ac (227)															

Lanthanides	$_{57}$La 138·9	$_{58}$Ce 140·1	$_{59}$Pr 140·9	$_{60}$Nd 144·2	$_{61}$Pm (147)	$_{62}$Sm 150·4	$_{63}$Eu 152·0	$_{64}$Gd 157·3	$_{65}$Tb 158·9	$_{66}$Dy 162·5	$_{67}$Ho 164·9	$_{68}$Er 167·3	$_{69}$Tm 168·9	$_{70}$Yb 173·0	$_{71}$Lu 175·0
Actinides	$_{89}$Ac (227)	$_{90}$Th 232·0	$_{91}$Pa 231·0	$_{92}$U 238·0	$_{93}$Np 237·0	$_{94}$Pu (242)	$_{95}$Am (243)	$_{96}$Cm (248)	$_{97}$Bk (247)	$_{98}$Cf (251)	$_{99}$Es (254)	$_{100}$Fm (253)	$_{101}$Md (256)	$_{102}$No (254)	$_{103}$Lw (257)

SI units

Physical quantity	*Old unit*	*Value in SI units*
energy	calorie (thermochemical)	4·184 J (joule)
	*electronvolt—eV	$1{\cdot}602 \times 10^{-19}$ J
	*electronvolt per molecule	96·48 kJ mol^{-1}
	erg	10^{-7} J
	*wave number—cm^{-1}	$1{\cdot}986 \times 10^{-23}$ J
entropy (S)	eu = cal g^{-1} °C^{-1}	4184 J kg^{-1} K^{-1}
force	dyne	10^{-5} N (newton)
pressure (P)	atmosphere	$1{\cdot}013 \times 10^5$ Pa (pascal), or N m^{-2}
	torr = mmHg	133·3 Pa
dipole moment (μ)	debye—D	$3{\cdot}334 \times 10^{-30}$ C m
magnetic flux density (H)	*gauss—G	10^{-4} T (tesla)
frequency (ν)	cycle per second	1 Hz (hertz)
relative permittivity (ε)	dielectric constant	1
temperature (T)	*°C and °K	1 K (kelvin); 0 °C = 273·2 K

(* indicates permitted non-SI unit)

Multiples of the base units are illustrated by length

fraction	10^9	10^6	10^3	1	(10^{-2})	10^{-3}	10^{-6}	10^{-9}	(10^{-10})	10^{-12}
prefix	giga-	mega-	kilo-	metre	(centi-)	milli-	micro-	nano-	(*ångstrom)	pico-
unit	Gm	Mm	km	m	(cm)	mm	μm	nm	(*Å)	pm

The fundamental constants

Avogadro constant	L or N_A	$6{\cdot}022 \times 10^{23}$ mol^{-1}
Bohr magneton	μ_B	$9{\cdot}274 \times 10^{-24}$ J T^{-1}
Bohr radius	a_0	$5{\cdot}292 \times 10^{-11}$ m
Boltzmann constant	k	$1{\cdot}381 \times 10^{-23}$ J K^{-1}
charge of a proton	e	$1{\cdot}602 \times 10^{-19}$ C
(charge of an electron = $-e$)		
Faraday constant	F	$9{\cdot}649 \times 10^4$ C mol^{-1}
gas constant	R	8·314 J K^{-1} mol^{-1}
nuclear magneton	μ_N	$5{\cdot}051 \times 10^{-27}$ J T^{-1}
permeability of a vacuum	μ_0	$4\pi \times 10^{-7}$ H m^{-1} or N A^{-2}
permittivity of a vacuum	ε_0	$8{\cdot}854 \times 10^{-12}$ F m^{-1}
Planck constant	h	$6{\cdot}626 \times 10^{-34}$ J s
(Planck constant)/2π	$\hbar$	$1{\cdot}055 \times 10^{-34}$ J s
rest mass of electron	m_e	$9{\cdot}110 \times 10^{-31}$ kg
rest mass of proton	m_p	$1{\cdot}673 \times 10^{-27}$ kg
speed of light in a vacuum	c	$2{\cdot}998 \times 10^8$ m s^{-1}

$\ln 10 = 2{\cdot}303$ $\quad \ln x = 2{\cdot}303 \lg x$ $\quad \lg e = 0{\cdot}4343$ $\quad \pi = 3{\cdot}142$

$R \ln 10 = 19{\cdot}14$ J K^{-1} mol^{-1} $\quad RTF^{-1} \ln 10 = 59{\cdot}16$ mV at 298·2 K